Math is Fun!
A guide to number sense and mental arithmetic

Tom Ferguson

DEDICATION

For three of my favorite math teachers Mrs. Ilene Smith, Mr. C.C. Turner, and Mr. Bob Wylie

Chapter 0: The Beginning 7

Chapter 1: Basics 8

Chapter 2: About this Book 9

Chapter 3 The Art of Number Sense 11

3.1 When to skip a problem… 11

3.2 Make sure you answer the question! 11

3.3 How to learn the tricks... Practice! 11

Chapter 4: Math Basics (EMH) 12

Chapter 5: Roman Numerals (EMH) 16

Chapter 6: Base Arithmetic (MH) 18

6.1 Introduction 18

6.2 How Base Numbering Schemes Work 18

6.3 Addition and Subtraction 19

6.4 Conversion between Bases 20

6.5 Conversion between bases involving powers of 2 21

6.6 Is there a 'base 1'? 23

Chapter 7: Time and Calendar (E) 24

Chapter 8 Subtraction (EMH) 26

8.1 Subtraction Through Addition (EMH) 26

8.2 Subtracting Reverses 26

Chapter 9: Multiplication (EMH) 27

9.1 Multiplying by Powers of 10   (EMH) 27

9.2 Using Common Fractions to Simplify Multiplication 27

9.3 Multiplying by 11 (EMH) 31

9.5 Multiplying by 111 (MH) 31

9.6 Multiplying by 101 (EMH) 32

9.7 Multiplying by 143 and its multiples(MH) 32

9.8 Multiplying by 167 (MH) 33

9.9 Multiplying by 3367 (EMH) 33

9.11 Multiplying by Numbers Near 100 (or any power of 10) 35

9.12 Multiplication by Factoring and Rearranging 36

9.13 Multiplying by 17 36

9.14 Multiplying Numbers Near 100 (MH) 37

9.15 Other Special Numbers 38

Chapter 10 Squaring Numbers that End in 5 and Related Tricks (EMH) 40

Chapter 11: Extending the Multiplication Tricks 42

11.1 Moving The Decimal 42

11.2 Using the Distributive Property 42

11.3 Using the Distributive Property in Reverse (MH) 43

Chapter 12 Multiplying Numbers That End in 5 (MH) 44

Chapter 13 Approximation Problems (EMH) 47

13.1 Using Multiplication Tricks (EMH) 47

13.2 Approximating Roots (MH) 48

13.3 Other Estimation Problems (H) 49

Chapter 14 Operations with Fractions (EMH) 51

14.1 Add/Subtracting Using the Cross Product 51

14.2 Comparison Using Cross Product 52

Chapter 15: Fraction and Mixed Number Tricks (MH) 54

15.1 Mixed Number Multiplication – whole part the same; fractional part adds to 1 54

15.2 Eliminating the Fraction 55

15.3 Shortcut for $(a \cdot a/b)$ 56

Chapter 16 Special Fraction Tricks (MH) 58

Chapter 17 Remainders and Modulo Arithmetic (EMH) 59

17.1 Remainder Tricks (EMH) 59

17.2 Modulo Arithmetic (MH) 61

Chapter 18 Common Fractions, Decimal, Percent (EMH) 63

Chapter 19 Set Theory (MH) 65

Chapter 20 Difference of Squares (MH) 67

Chapter 21 More Squares Tricks (H) 68

21.1 $x^2 + y^2 - (x-y)^2$ 68

21.2 $x^2 - (x-1)^2 + (x-2)^2 + (x-3)^2$ etc. 68

21.3 $(x-n)^2 + (x+n)^2$ 69

21.4 $x^2 + (x+1)^2$ 69

21.4 $x^2 + (x+n)^2$ 70

Chapter 22 Product of 4 Consecutive Integers +1 (H) 72

Chapter 23 Sequences and Series (MH) 73

23.1 Arithmetic Series 73
23.2 Geometric Series 75
23.3 Infinite Geometric Series 76
23.5 Sum of Short Series 77
Chapter 24 Repeating Decimals (MH) 78
24.1 Introduction 78
24.2 All Digits Repeat 78
24.3 Not all Digits Repeat 79
Chapter 25 Median, Average, Mode (EMH) 81
25.1 Average 81
25.2 Median 81
25.3 Mode 82
Chapter 26 Geometry (MH) 83
26.1 Formulae and Definitions for Geometry 83
26.2 Polygons 86
26.3 Pythagorean Theorem 88
26.4 Geometric Mean 89
26.5 Analytic Geometry 90
26.6 Polar Coordinates (H) 91
Chapter 27: Trigonometry (H) 94
27.1 Basic Trig Notation and Values 94
27.2 Pythagorean Identities 95
27.3 Additive Identities 96
27.4 Double Angle Formulas 96
27.5 Period 97
27.6 Sample Problems 97
Chapter 28 Squaring Numbers Near 50 (MH) 99
Chapter 29 Elementary Number Theory (MH) 100
29.1 Number of Positive Integral Divisors 100
29.2 Sum of Positive Integral Divisors 101
29.3 Greatest Common Factor and Least Common Multiple (MH) . 101
29.4 The Concept of Relatively Prime 103
29.5 The Even Prime 103
29.6 Triangular numbers, including sums thereof! 104
29.7 Pentagonal Numbers 105

29.8 Octagonal Numbers 105

Chapter 30 Putting Radicals In Their Place! (H) 107

30.1 Basic Principles 107

30.2 Difference of Squares with Radicals 108

Chapter 31 Complex Numbers (H) 110

Chapter 32: Matrix Theory 112

Chapter 33 Basic Algebra (MH) 114

33.1 Some Basics 114

33.2 Ratio and proportion 115

33.3 Evaluating a function 116

33.4 Inequalities 117

33.5 Quadratic equations/formula 117

33.6 The sum and product of roots for quadratic equations 118

33.7 Product of roots for cubic equation 118

33.8 Inverse functions 119

33.9 Polynomial Remainders 120

Chapter 34 Exponents and Logarithms 120

34.1 Laws of Exponents 120

34.2 Logarithms – Definition and Notation 123

34.3 Properties of Logarithms – Why the Slide Rule Died 123

Chapter 35 Probability and Odds (EMH) 127

35.1 Probability 127

35.2 Table of Probabilities for Two Dice (MH) 130

Chapter 36 Counting Problems – Combinations and Permutations (H) 131

36.1 Factorials – It's Exciting! (H) 131

36.2 Combinations: "n Choose k" problems 132

36.3 Permutations 133

Chapter 37 Binomial Expansion 134

Chapter 38: Calculus - It Rocks! (H) 137

38.1 Limits 137

38.2 Vertical Asymptotes 138

38.3 Derivative of Polynomials 139

38.4 Integrals of Polynomials 140

38.5 Min/Max Values of a Function 142

Appendix A Conversions (EMH) 144
  A.1 Metric prefixes: 144
  A.2 common conversions 144
  x.3 Fahrenheit/Centigrade Conversion 145
Appendix B Powers   (EMH) 146

Chapter 0: The Beginning

When I was in 6th grade at Judson Jr. High (located in Judson, Texas!), a senior from nearby Longview High School came to tell us about the UIL math competitions, which at that time was only for high school students and consisted of two contests: Slide Rule and Number Sense. I was hooked on the idea of "mental math" immediately. It was four years until I got to act on my interest. In spite of my confidence that I would be great at this, the first year did not go well. I tried my hand at Slide Rule first; the results were terrible. Towards the end of the year, I decided to try Number Sense as well –maybe I was good at that! At the last league competition of the year, I competed in both and scored a -5 in Number Sense and some forgettable number in Slide Rule. Most people would have given up by then, but not me. I practiced hard that summer; learning and rehearsing every trick. By the time school started, I was ready. I took first place in Number Sense at the first competition of the year. The next two years were very successful for me as I qualified for the UIL State Meet both years, finishing 7th as junior and 4th as a senior. I also figured out what I was doing wrong in Slide Rule and went to state in that as well; finishing 4th as a senior.

So where did this book come from? Back in high school, I began collecting the tricks in handwritten forms in a notebook we all simply referred to as "The Book". Anyone interested in Number Sense at our school would borrow "The Book" and make their own handwritten copy. My friend Tom George made such a copy and his actually survived to the present day. He graciously gave me a copy back in 2002 and I began contemplating modernizing our collection of tricks and putting it in a useful up-to-date form. After working with the Hebron High and Polser Elementary UIL math teams over the last couple of years, I finally got the inspiration to finish the project. Using the original tricks from back in the 70's as a base, I also studied current number sense tests and learned the new tricks (see, it is possible, despite what they say about old dogs!) Of course, a lot of number sense is not tricks, but just knowing some math. I put plenty of that in here too.

To me, math and mental math are fun. I hope you read this and catch some of that fun spirit!

Chapter 1: Basics

Number sense is an exciting way to learn mathematical principles and develop your ability to apply those principles quickly and accurately. A competition test consists of 80 problems with a 10-minute time limit – that is only 7.5 seconds per problem!

Key elements to success:

- Rapid problem recognition: What problem is being solved?
- Knowledge of arithmetic facts (for example, 2+2=4) for addition, subtraction, multiplication, and division
- Knowledge of basic math principles (the Distributive property, for example)
- Commitment to practice in order to develop mental computational ability, number recall, and speed
- Ability to write quickly and clearly – if your answer can't be read, it's wrong!

Chapter 2: About this Book

If you got assigned to work with the number sense team and do not know where to start, hopefully this book will help you out! This book is probably a little heavy to drop on the average elementary student but an ambitious middle schooler and certainly high school students can tackle most of these subjects on their own.

The scope of this work is not intended to be exhaustive. However, I have tried to cover every trick, both past and present, that I see arising on Number Sense tests. In addition to tricks, there is math to be learned that the student may not have encountered yet (e.g., Trig, Calculus). Here I try to provide enough background so that the student can learn enough to solve the problems that they may confront in Number Sense.

Speed benchmarks: To really be successful at Number Sense, you have to be fast. The way to get fast is to practice – a lot! I thought it would be helpful to give an indication of how fast is "fast". To that end, I have provided what I call "speed benchmarks" along the way. These are 5 problem tests that the reader can use to measure their speed progress. The benchmarks are subjective except for the fastest category; here I have measured my own speed and scored that as "Excellent".

EMH – These letters, or some subset thereof, appear in the titles of many of the sections. These are intended to give a rough guide to the level of test to which the chapter or section would apply. E – Elementary M- Middle School H- High School

MATH IS FUN!

Chapter 3 The Art of Number Sense

3.1 When to skip a problem…

If you are going to excel at Number Sense, you goal must be finishing the test in 10 minutes or less. As we noted before, that is 7.5 seconds per problem. Of course, some problems will take more and others less. Some problems are time consuming, especially if you do not know the trick, that you are tempted to skip them.

So when do you skip a problem? Skipping a problem means taking the 4 point hit for an incorrect answer, so it is not to be done lightly. However, if you come to a problem you do not know how to solve, but you see the next one is easy, then it makes sense to skip the first one. This nets you one point and keeps you moving. If you confront two or three problems in a row that you are tempted to skip because you do not know how to solve them quickly, this probably means you are out of your comfort zone on the test, i.e., you are in new territory with some unfamiliar problems.

3.2 Make sure you answer the question!

I'll say it again: make sure you answer the question! If the answer calls for "cents", putting your answer in dollars is going to be wrong.

3.3 How to learn the tricks… Practice!

The way to get good at math tricks is the same way you get good at anything else:

Practice!

Practice tips:

1) When you learn a new trick, do specific practice on the trick until you have it down cold.
2) Make up your own tests. Writing good tests makes you think about how the tricks work. Trade tests with a friend.
3) Take practice tests and record your results (problems worked, score, errors made). Analyze your speed progress; also take note of your

errors. What do you need to improve on? Which ones did you get right, but took too long to do?

4) Practice doing multiplication in your head and check your answer with a calculator. See how many partial answers you can juggle in your head at the same time. Example: Try multiplying 138 x 138 x 138 without writing anything down but the answer.

5) Set a goal for how many practice tests you will take per week and make yourself accountable for it.

Chapter 4: Math Basics (EMH)

A few odds and ends before we get started.

1) Know the basic arithmetic facts (9 +12=21; 7 x 8=56; 64/8 = 8, etc). This especially applies to the elementary number sense contestants who are just mastering multiplication and long division. Use flash cards or whatever it takes to make sure you know the basics.

2) Order of operations: In problems without parentheses, division and multiplication take precedence over addition and subtraction.

3) A palindrome is a word or phrase that reads the same backwards and forward: ("A man, a plan, a canal – Panama!"). In Number Sense, a palindrome is a number that read the same forwards and backwards (example: 373).

4) The additive inverse of a number (also referred to as 'the negative' of a number), when added to the number, yields 0. Thus -7 is the additive inverse of 7.

5) The reciprocal (also referred to as the 'multiplicative inverse") of a number, when multiplied by the number, yields 1. Thus, ½ is the reciprocal of 2.

6) A negative times a negative is a positive. A positive times a negative is a negative.

7) Number Sense rules are pretty narrow regarding extra 0's. Do not put them in there! For example, if the answer is .17, then 0.17 is not acceptable and neither is .170.

8) There are five commonly used ways to indicate multiplication: x, $\cdot$, *, the word "times", and juxtaposition (i.e., placing two things next to each other). On number sense tests, the "x" tends to be most often used. I have used the "$\cdot$" symbol most often in this book.

A theme throughout the number sense tricks:

It is easier and faster to do two steps (or more!) than one. We will go into each of these tricks in more detail   throughout the book.

Example: Consider the following problem:

$$235 - 97$$

The obvious way to do this is to mentally execute subtraction just as you would on paper, doing the borrowing and   so forth. However, visualize the problem like this:

$$235 - (100\text{-}3) \text{ and it becomes an easy two-step problem!}$$

$$135 + 3 = 138$$

Example: $298 + 624 =$

$$= 300 + 622$$

$$= 922$$

Here we simply added 2 to 298 and subtracted the same from 624 resulting in a much simpler problem.

Example: $35 \cdot 12$

We can determine the answer to this by multiplying 35 by 2 (to get 70) and dividing 12 by 2 (to get 6):

$$35 \cdot 12 \ = 70 \cdot 6 = 420$$

Example 4: $5 \, \tfrac{1}{2} \cdot 68 = 11 \cdot 34 = 374$

Example 5: $3 \, 1/3 \cdot 42 = 10 \cdot 14 = 140$

Speed benchmark: Time your speed on these five problems:

1. $38 + 97 =$ _________

2. $55 \cdot 32 =$ ________
3. $451 - 297 =$ _______
4. $652 + 587 =$ _______
5. $4 \, \tfrac{1}{2} \cdot 38 =$ _________

Speed Rating

< 8 seconds: Excellent

8-11 seconds: Good

12-20 seconds: Average

21-   seconds: Needs improvement

Chapter 5: Roman Numerals (EMH)

A number sense test is usually not complete without a problem involving Roman numerals. There are not any great tricks for this; you just have to know the symbols and the three basic rules.

Basic Rules:

1)   No more than 3 in a row of one symbol
2)   Small in front of large: Subtract (Example: IV = 4)
3)   Small after large: Add (Example: VI = 6)

1	I
5	V
10	X
50	L
100	C
500	D
1000	M

Example: Express 42 in Roman Numerals: XXX = 30, and three is the limit, so we have to go to 50 (which is L) and subtract 10 (by putting X in front of L):

XL = 40

Now we need 2, so we add II; so 42 = XLII

Example: Convert CLXIX to Arabic numerals: C is 100, L is 50, the first X is 10, so this adds to

160; now we have an I before X, which means subtract 1 from 10 leaving 9; so

CLXIX = 169.

Chapter 6: Base Arithmetic (MH)

6.1 Introduction

Probably due to the fact that we humans have 10 fingers (except for my Aunt Lucy, who had 12, but that's a different story), we use base 10 as our numbering system. This means that each place in a number represents a power of 10. These are often referred to as the "tens place", "hundred place", "thousands place", etc. The base of a number is indicated by a subscript (a number below the line). Base 10 is implied when no subscript is given.

Arithmetic in base 2 and base 16, and to a lesser extent base 8, can be important in describing computer science concepts. Note that we also use a modified form of base 60 for keeping time. The abacus, an ancient calculating device, used base 4 arithmetic. The rest of the base problems you encounter as number sense problems are just that – number sense problems!

6.2 How Base Numbering Schemes Work

Base arithmetic is not hard to understand once you grasp the "base-ic" concept. This can be illustrated by writing out number as an expression of powers of 10:

$$2561 = 2 \times 10^3 + 5 \times 10^2 + 6 \times 10^1 + 1 \times 10^0$$

(Note that $10^0 = 1$, for that matter, any number to the zero power $=1$, except 0 to the 0, which is undefined.)

Now consider a number written in base 6:

$$3254_6 = 3 \times 6^3 + 2 \times 6^2 + 5 \times 6^1 + 4 \times 6^0$$

One of the first things to notice is that the digit 6 (or for that matter, 7, 8, or 9) will not appear in a number written in base 6. This is because 6 is represented as 10_6.

Compare the two rows of the following table with the top row being base 10 and the second row being base 6:

1	2	3	4	5	6	7	8	9	10	11	12	13	14	15	16	17	18
1	2	3	4	5	10	11	12	13	14	15	20	21	22	23	24	25	30

Note that $20_6 = 12$ because $20_6 = 2 \times 6^1 + 0 \times 6^0$, and so forth.

6.3 Addition and Subtraction

Now we can get to some number sense problems. The first problems we will consider call for adding and subtracting within a base. These are done exactly like base 10 addition and subtraction except you have to remember what you are "borrowing" and "carrying" is a power of the base specified, not base 10! If there is no borrowing or carrying, then the problem is very easy.

Example: $234_7 - 123_7 = 111_7$

As you can see this is an easy example. 4-3 yields 1 in the ones place, 3-2 yields 1 in the 7's place; and 2-1 yields 1 in the 7^2 place.

Example: $321_6 - 253_6 =$ _________$_6$

This one is a little harder since we are going to have to borrow.

Let's walk through it step by step:

1-3, we need to borrow from the 2; what we are borrowing is a '6', so we get 7-

3=4 as our ones digit.

Next, we have 1 –5, so we borrow again (once again, we are borrowing a '6'), so we get 7-5=2.

Finally, we have 2-2=0. So our answer is

$$321_6 - 253_6 = 24_6$$

CAUTION! Be sure to write your answer in the specified base! This is a specific application of the general number sense rule to always answer the question. For example, the problem may be in base 6 but the answer called for in base 10.

Example: $428_9 + 372_9 = $ _______$_9$

We proceed as we would with any addition problem; 8+2 = 10; which is 9+1, so we write down 1 for the '1s' digit and carry 1 (the 9 is a 1 in the 9's place. Confusing, huh?). Next,we add 2 + 7 + 1 (from the carry) and we get 10 again. So we write 1 in the 9's place and carry 1 again. Now we have 4 + 3 + 1 (from the carry) which is 8, so our answer is:

$$811_9$$

6.4 Conversion between Bases

Conversion between bases may first call for doing addition or subtraction.

Example: $11_4 + 23_4 = $ ____ $_{10}$

Note that the answer is called for in base 10, so we cannot write our answer down as we go along. You will have to keep the answer in your head, then do the conversion before writing down the final answer.

1 +3 = 4, so we remember 0 and carry 1, 1 + 2 + 1 = 4, so we remember 0 and carry 1. The answer we have in our head now is 100_4

To convert this to base 10, we multiply the 1 by 4^2, since that is the place it represents, to get 16. The other two places being 0, we are done.

$$11_4 + 23_4 = 16_{10}$$

Example: $55_{10} = $ _____ $_8$

We have to ask what is the highest power of 8 that will divide into 55 and how many times. The $8^2 = 64$, so that is too big. So we use 8, which will divide in 6 times giving 48 and leaving 7. So we have 6 in the 8's place and 7 in the 1's place; our answer is 67_8.

6.5 Conversion between bases involving powers of 2

The following table is given to illustrate the pattern that develops in converting between bases involving the power of 2:

Base 10	Base 2 (2^1)	Base 4 (2^2)	Base 8 (2^3)	Base 16 (2^4)
1	1	1	1	1
2	10	2	2	2
3	11	3	3	3
4	100	10	4	4
5	101	11	5	5
6	110	12	6	6
7	111	13	7	7
8	1000	20	10	8
9	1001	21	11	9
10	1010	22	12	A
11	1011	23	13	B

12	1100	30	14	C
13	1101	31	15	D
14	1110	32	16	E
15	1111	33	17	F
16	10000	40	20	10

(Note: The letters A-F are used in base 16 to represent the numbers 10-15. Number sense problems have not used base 16, it is included here for completeness).

Conversions between from base 4,8, or 16 to base 2 are a matter of considering the digits grouped by 2,3, or 4 digits at a time. For base 4, the digits are grouped by 2, for base 8, by 3, and for base 16, by 4. Study the table carefully and you will see the pattern!

Example: $201_4 = $ _______________ $_2$

See our table above, each digit in base 4 becomes 2 digits in base 2. 2 = '10', 0 = '00' and 1 = '01'.
  Therefore,

  $201_4 = 100001_2$

Example: $62_8 = $ _______________ $_2$

See our table above, each digit in base 8 becomes 3 digits in base 2. 6 = '110', 2 = '010'

Therefore,

  $62_8 = 110010_2$

The same principle works in reverse:

Example: $10111011_2 =$ _________________ $_8$

Start from the right, take 3 digits at a time and convert to base 8, then write down your answer.

'011' = 3; '111' = 7, '010' = 2

Therefore,

$10111011_2 = 273_8$

6.6 Is there a 'base 1'?

You had to ask! Yes, there is base 1 and it is actually used in real life, although I have never seen a number sense problem involving it. Base 1 is simply 'tally marks'. As in, 4 = | | | |, and 5 is represented by drawing a diagonal line across the 4 marks, and so forth. This was probably the first base numbering system!

Chapter 7: Time and Calendar (E)

Elementary tests include questions that are pretty easy if you know the basic time and calendar facts.

Month	Number of Days
January	31
February	28 (or 29 in leap year)
March	31
April	30
May	31
June	30
July	31
August	31
September	30
October	31
November	30
December	31

There is a poem for remembering this: *"30 days hath September/April, June and November/All the rest have 31/Except February which has 28!"*
Of course, in a leap year, February has 29 days. Leap years are those years divisible by 4, except that a year ending in '00' is not a leap year unless it is also divisible by 400.

From	To
60 seconds	1 minute
60 minutes	1 hour
24 hours	1 day
7 days	1 week
365 (or 366 in a leap year)	1 year

Example: 5 weeks = __________ days

$5 \cdot 7 = 35$

Example: How many seconds are in 2 minutes?

$2 \cdot 60 = 120$

Example: How many days are in June?

Just remember the table! 30

Chapter 8 Subtraction (EMH)

8.1 Subtraction Through Addition (EMH)

When the subtrahend (that's the second term in the subtraction equation; the first is known as the minuend) is near a multiple of 100, 1000, etc, it is often easier to change the problem as indicated in the following example:

Example: $642 - 297 =$

$$= 642 - 300 + 3$$
$$= 342 + 3$$

$$= 345$$

8.2 Subtracting Reverses

When the two 3 digit numbers being subtracted are 'reverses' of each other, that is, if you read one backwards you get the other, the difference can be found quite easily!

Example: $783 - 387 =$

To find the answers, subtract the hundreds digits $(7-3 = 4)$ and multiply by 99:

$$4 \cdot 99 = 396$$

And you are done!

Chapter 9: Multiplication (EMH)

9.1 Multiplying by Powers of 10   (EMH)

Multiplying by 10 or any power of 10 is as simple as adding zeroes.

Example: $15 \cdot 10$

The answer is as simple as adding a zero to 15 to get 150.

Example: $283 \cdot 100 = 28300$

Just add two zeroes!

Do you get the pattern? Good! This basic fact is necessary to use the common fraction tricks.

9.2 Using Common Fractions to Simplify Multiplication

The common fraction tricks open up a lot of possibilities to make multiplication easier.

One of the basics of elementary number sense is 25. The trick is to recognize that 25 is equal to $\dfrac{100}{4}$. Therefore, multiplying by 25 is the same as dividing by 4, then multiplying by 100 (which is quite easy, as we saw in the previous section).

Example: $25 \cdot 64 =$

$$\frac{100}{4} \cdot 64 \ = 100 \cdot 16 = 1600$$

Example: $204 \cdot 25 = 204 \cdot \dfrac{100}{4} = 51 \cdot 100 = 5100$

Now try a tougher problem:

Example: $375 \cdot 168 = \dfrac{3000}{8} \cdot 168 = 3000 \cdot 21 = 63000$

At the elementary level, the tests will usually stick with multiplying by 25 and 125. Everyone else should be prepared for anything!

Speed benchmark: Time yourself on using the 25 trick:

1. $25 \cdot 48 =$ _______	Excellent: < 8 seconds
2. $25 \cdot 84 =$ _______	Good: 8- 12 seconds
3. $72 \cdot 25 =$ _______	Average: 13-20 seconds
4. $104 \cdot 25 =$ _____	Beginner: 20-45 seconds
5. $308 \cdot 25 =$ _____	Keep working!: > 45 seconds

Using Common Fractions to Multiply

To Multiply by:	Use the Fraction:
5	$\dfrac{10}{2}$
12 ½	$\dfrac{100}{8}$
16 2/3	$\dfrac{100}{6}$
25	$\dfrac{100}{4}$
33 1/3	$\dfrac{100}{3}$
37 ½	$\dfrac{300}{8}$
50	$\dfrac{100}{2}$
62 ½	$\dfrac{500}{8}$
66 2/3	$\dfrac{200}{3}$

75	$\dfrac{300}{4}$
87 ½	$\dfrac{700}{8}$
125	$\dfrac{1000}{8}$
250	$\dfrac{1000}{4}$
375	$\dfrac{3000}{8}$
500	$\dfrac{1000}{2}$
625	$\dfrac{5000}{8}$
750	$\dfrac{3000}{4}$
875	$\dfrac{7000}{8}$

More examples:

Example: $62 \; ½ \cdot 128 = \dfrac{5000}{8} \cdot 128 = 5000 \cdot 16 = 80000$

Example: $33\dfrac{1}{3} \cdot 156 = \dfrac{100}{3} \cdot 156 = 100 \cdot 52 = 5200$

9.3 Multiplying by 11 (EMH)

Example:

$$134 \cdot 11 = 1474$$

Imagine 134 as 0134.0   Starting from the right, add each number to its neighbor and write down the result, beginning in the 1's place. Remember to carry when necessary!

9.4 Multiplying by 51 (MH)

This trick applies to multiplying two digit even numbers by 51. The trick is two simple steps:

- Take half of the two digit number; write it down
- Write down the two digit number.

Example:

$$51 \cdot 22 = (1/2 \text{ of } 22) \text{ append } 22 = 1122$$

9.5 Multiplying by 111 (MH)

The '11' trick can be extended to multiply by 111:

465 · 111   Imagine 465 as 00465.00. Starting from the right, add 3 digits and right down the answer beginning in the 1's place (but remember to carry!)

$$1\text{'s} = 0 + 0 + 5 = 5$$

$$10\text{'s} = 0 + 5 + 6 = 1 + \text{carry } 1$$

$$100\text{'s} = 5 + 6 + 4 + \text{carry } 1 = 6 + \text{carry } 1$$

$$1000\text{'s} = 6 + 4 + 0 + \text{carry } 1 = 1 + \text{carry } 1$$

$$10000\text{'s} = 4 + 0 + 0 + \text{carry } 1 = 5$$

Answer: 51615

9.6 Multiplying by 101 (EMH)

This is an easy one: Two multiply a two digit number by 101, you just write the two digit number twice!

Example:

$24 \cdot 101 = 2424$

9.7 Multiplying by 143 and its multiples(MH)

The key fact here is that $7 \cdot 143 = 1001$; the problem will be easy to solve if the number you are multiplying by 143 is a multiple of 7, since you will be able to divide 7 out of that factor and multiply it by 143 to make the problem easy.

Multiplying by multiples of 143, namely, 286, 429, 572,715, and 858 work in a similar fashion:

Example: $28 \cdot 143 = $ ______________

$$28 \cdot 143 = 28 \cdot \frac{1001}{7} = 4 \cdot 1001 = 4004$$

Example: $119 \cdot 429 = $ __________

$$119 \cdot \frac{3}{7} \cdot 1001 = 17 \cdot 3 \cdot 1001 = 51 \cdot 1001 = 51051$$

9.8 Multiplying by 167 (MH)

Ever notice that $6 \cdot 167 = 1002$?

Example: $36 \cdot 167 = 36 \cdot \frac{1002}{6} = 6 \cdot 1002 = 6012$

9.9 Multiplying by 3367 (EMH)

You are probably getting the pattern by now: $3 \cdot 3367 = 10001$; just as with the 143 trick, the problem will be easy to solve if the number you are multiplying is a multiple of 3.

Example:

$$243 \cdot 3367 = (81 \cdot 3) \cdot 3367 = 81 \cdot 10001 = 810081$$

9.10 Multiplying by 14443 (MH)

Ok, you are never going to guess how this trick works... or maybe you have the concept by now!

Note that $7 \cdot 14443 = 101101$

Example: $63 \cdot 14443 =$

$$9 \cdot 7 \cdot 14443 =$$

$$9 \cdot 101101 =$$

$$909909$$

Practice:

1. $52 \cdot 101 =$ _______________
2. $56 \cdot 14443 =$ _______________
3. $175 \cdot 429 =$ _______________
4. $167 \cdot 306 =$ _______________
5. $87 \cdot 3367 =$ _______________

Speed benchmark: Time yourself on using the 25 trick:

<table>
<tr><td>6. 25 · 48 = ______</td><td>Excellent:</td><td>< 20 seconds</td></tr>
<tr><td>7. 25 · 84 = ______</td><td>Good:</td><td>21-30 seconds</td></tr>
<tr><td>8. 72 · 25 = ______</td><td>Average:</td><td>31-45 seconds</td></tr>
<tr><td>9. 104 · 25 = ____</td><td>Beginner:</td><td>46-60 seconds</td></tr>
<tr><td>10. 308 · 25 = ____</td><td>Keep working!:</td><td>> 60 seconds</td></tr>
</table>

9.11 Multiplying by Numbers Near 100 (or any power of 10)

Multiplying by numbers that are close to 100 is easy using the distributive property. The same trick applies for any power of 10.

Example: 52 x 103 =

$$= 52 \times (100 + 3)$$

$$= 52 \times 100 + 52 \times 3$$

$$= 5200 + 156$$

$$= 5356$$

Example: 71 x 997 =

$$= 71 \times (1000 - 3)$$

$$= 71 \times 1000 - 71 \times 3$$

$$= 71000 - 213$$

$$= 70787$$

9.12 Multiplication by Factoring and Rearranging

This trick relies on vision to see how to break the problem apart to make it easier, as in, you are going to read the example and say "No one could ever solve that in 7.5 seconds!". It bears looking at anyway since it is an important concept.

Example: $412 \cdot 187$

$$412 \times 187 = 4 \times 103 \times 17 \times 11 = (4 \times 17) \times 103 \times 11 = 68 \times 103 \times 11$$

We have now made the problem much easier!

$$68 \times 103 = 68 \times (100 + 3) = 6800 + 204 = 7004$$

Now all we have left is to compute $7004 \times 11 = 77044$

9.13 Multiplying by 17

When you see 17 lurking in a multiplication problem (as it was in the prior example as a 'red herring'), the first thing to look for is whether there is also a factor of 6 in the equation. This is because $6 \times 17 = 102$.

Example: $552 \times 85 =$

$$= 92 \times 6 \times 17 \times 5$$

$$= 92 \times 102 \times 5$$

$$=460 \times 102$$

$$= 46920$$

9.14 Multiplying Numbers Near 100 (MH)

There are several tricks available for multiplying numbers near 100; the first trick is to quickly recognize which trick is going to be easiest to execute!

The general purpose trick is to apply the formula:

$$(x + a)(x + b) = x^2 + (a+b)x + ab$$

In the cases we consider, x will always be 100, so $x^2 = 10,000$; this makes most of the arithmetic involved easy.

Example 1: $93 \cdot 97 = (100-7)(100-3)$

$$= 10000 - (7+3) \cdot 100 + 21$$

$$= 10000 - 1000 + 21$$

$$= 9021$$

Example 2: $96 \cdot 109 = (100-4)(100+9)$

$$= 10000 + (9-4) \cdot 100 - 36 \quad \text{(Watch your signs!)}$$

$$= 10464$$

Example 3: $97 \cdot 103 = (100-3)(100+3)$ In this case, use the difference of squares!

$$= 10000 - 9$$

$$= 9991$$

Example 4: $94 \cdot 101 = 9494$ Use the trick for multiplying a two digit number by 101 (i.e., write the two digit number twice).

Example 5: $93 \cdot 99 =$ Sometimes, it is easier just to use the Distributive Property!

$$= 93 \cdot (100 - 1)$$

$$= 9300 - 93$$

$$= 9207$$

Speed benchmark: Complete these 5 problems in less than 25 seconds:

1.   95 x 105 = ______________
2.   91 x 94 = ______________
3.   98 x 106 = ______________
4.   93 x 101 = ______________
5.   102 x 108 = ______________

9.15 Other Special Numbers

Note that $37 \cdot 27 = 999$

$$69 \cdot 29 = 2001$$

Chapter 10 Squaring Numbers that End in 5 and Related Tricks (EMH)

This is one of my favorite tricks! Note that "squaring a number" mean to multiply it by itself. To square the number 75 can be written in two different ways:

$$75^2 = 75 \cdot 75 = 5625$$

To easily compute this, take the ten's digit (7), add 1 (resulting in 8) and multiply these two numbers (7 · 8 =56). Then put 25 as the last two digits. It's that easy! It even works for larger numbers:

Example: 105 · 105 = 11025 (10 · 11= 110; then add on the 25)

This trick can be extended in a couple of interesting ways…

If the problem calls for multiplying two numbers where the digits are the same except for the 1's digits; and the 1's digits add up to 10; then the trick still works with a slight modification:

Example: 63 · 67 = 4221

> We still get the 42 from multiplying 6 · 7, the 21 comes from multiplying 3 · 7

> Another example:

> 91 · 99 = 9009   (9 · 10 gives the 90; 1 · 9 gives the '09'; note you need 2 digits!)

If the problem involves a 3 digit number that ends in two fives, you can use the trick twice, along with the Distributive Property, to get the answer:

Example: 255 · 255

The trick would be to multiply 25 · 26. Use the Distributive Property, this is the same as 25 · (25 +1) = 25· 25 + 25. Use the trick once to get 25 · 25 =625, add 25 to get 650. Now apply the 2nd '25' to get the final answer: 255 · 255 = 65025.

Ok, what about fractions? or decimals?

Example: $(4 \frac{1}{2})^2 = (4.5)^2$ What's the difference between these problems and squaring 45? Just the position of the decimal; our trick still works.

$(4.5)2 \quad = 20.25$

Example: $(35.5)^2$

Here we need to multiply 35 · 36 = 35 · (35 + 1) = 1225 + 35 = 1260; then we append .25 (instead of 25); so the answer is: 1260.25

Caution: Be sure to put your answer in the proper form – it may call specifically for a mixed number or a decimal!

Chapter 11: Extending the Multiplication Tricks

In a previous chapter, a table of multiplication tricks is given. These mostly consist of converting a whole number (for example, 25) to an improper fraction (100/4). This kind of trick can be extended to cover other cases by realizing that a change in the decimal place does not affect the basic trick.

11.1 Moving The Decimal

For example, just as above 25= 100/4, then 2.5 = 10/4, 250=1000/4, etc.

Also, 125=1000/8, 12.5 = 100/8, and 1.25 = 10/8

Example: 1.25 · 64 Divide 64 by 8 resulting in 8; multiply 10 to get 80.

11.2 Using the Distributive Property

Another simple extension is to use the Distributive Property (See Appendix A) to break a problem into two parts, one of which gives you a chance to use a trick.

Example 1: 26 · 48 = (25 + 1) · 48 = (25 · 48) + (1 · 48)

Now we can use the trick to compute 25 · 48 = 1200, and then just add 48, so the final answer is 1248

Example 2: 32 · 52 = (32 · 50) + (32 x 2)

We remember 50=100/2, so 32 · 50 = 1600; now we add 32 · 2 =64 to get the result of 1664.

11.3 Using the Distributive Property in Reverse (MH)

The distributive property can also be used in reverse when that makes the problem easier. This is a common number sense problem:

Example: $31 \cdot 17 + 31 \cdot 43 = 31 \cdot (17 + 43) = 1860$

Chapter 12 Multiplying Numbers That End in 5 (MH)

In a previous chapter, you learned how to square numbers that end in 5. This chapter extends that trick to multiply *any* two numbers that end in 5.

Example: Find $45 \cdot 85$

The trick works as follows: take the product of the tens digits $(4 \cdot 8 = 32)$; add the average of the tens digit $((4 + 8)/2 = 6)$ to get 38; then append 25 as the last two digits.

Answer: 3825

Example: Compute $135 \cdot 215$

$13 \cdot 21 = 273$; we add the average $(13 + 21)/2 = 17$ to get 290; then append the 25.

Answer: 29025

You may be thinking: if one of the tens digits is even and the other is odd, won't the average result in a fraction of ½? Yes! In this case, you will need to append '75' instead of '25' as the last two digits.

Example: Find $85 \cdot 175$

$8 \cdot 17 = 136$; the average is 12 ½ ; $136 + 12 = 148$; append the 75 (because of the ½).

Answer: 14875

Let's try one more of those 'odd' ones:

Example: Find $35 \cdot 65$

$3 \cdot 6 = 18$; $(3+6)/2 = 4$ ½ ; we $18 + 4$ to get 22; append 75 (because of the ½).

Answer: 2275

Why does this trick work? (H)

Skip this part if you are not curious yet or have no algebra background.

Express the problem as follows:

$(10x + 5)(10y + 5)$; these represent our two integers that end in 5 (assuming x and y are integers!).

Multiplying this out, we get:

$100xy + 50x + 50y + 25 =$

$100xy + 100 (x+y)/2 + 25 =$

$100 (xy + (x+y)/2) + 25$

So we see the 'xy' is the product of the tens digits and $(x+y)/2$ is the average of the tens digits. If there is a ½ left over after the average, then ½ · 100 = 50 added to the 25 gives us the 75. Voila!

Speed benchmark:

1. $85 \cdot 175 =$ _________
2. $35 \cdot 95 =$ _________
3. $65 \cdot 215 =$ _________
4. $75 \cdot 135 =$ _________
5. $55 \cdot 155 =$ _________

How did you do?

Excellent < 24 seconds

Good   24-30 seconds

Average   30-45 seconds

Beginner > 45 seconds

Chapter 13 Approximation Problems (EMH)

Every tenth number sense problem, beginning with number ten, is an approximation problem. This means that an answer will be deemed correct if it is within 5% of the exact answer. These problems are indicated by a "*" appearing before the problem number as follows:

Example: *20) $125 \cdot 241$

Note that the exact answer can be computed using the trick for multiplying by 125:

$$125 \cdot 241 = \frac{1000}{8} \cdot 241 = 1000 \cdot 30.125 = 30125$$

The problem will be counted correct if any answer between 28618 and 31631 is given.

13.1 Using Multiplication Tricks (EMH)

In the example above, the exact answer was determined using a common fraction trick.

However, the problem could have been simplified as follows:

$$125 \cdot 240 = \frac{1000}{8} \cdot 240 = 30000$$

Note that this would be been well within the accepted answer range and should be much faster to compute!

This example is a typical approximation problem at the elementary level.

13.2 Approximating Roots (MH)

Another typical approximation problem involves estimating the square or cube root of a large number. Here, it is helpful to remember two facts:

1) The square root of 100 is 10.
2) $\sqrt{ab} = \sqrt{a} \cdot \sqrt{b}$

First, we will compute a large square root to illustrate these two principles:

Example: Find the square root of 32400

$$\sqrt{32400} = \sqrt{324} \cdot \sqrt{100}$$

$$= 18 \cdot 10$$

$$= 180$$

Example: Approximate the square root of 155724

$$\sqrt{155724} = \sqrt{1557.24} \cdot \sqrt{100} \cdot$$

Assuming you have a lot of squares memorized, you remember that 39^2 = 1521 and 40^2 = 1600; so let's split the difference and guess that the square root of 1557.24 is about 39.5.

$$\text{Our estimate is thus } 39.5 \cdot 10 = 395$$

The answer is 394.61 so the accepted answer range would be 375-414. We did pretty well, huh?

So, the trick is to take out "100s" until you get the number into a range where you recognize the square root. Then, take your guess and multiply by 10 as many times as you took out 100.

What number would you take out to approximate cube roots? If you said 1000, you are correct!

Example: Approximate the cube root of 1386224

$1386224 = 1386 \cdot 1000$ From here, we note that $11^3 = 1331$ and $12^3 = 1728$. Since 1386 is pretty close to 1331, we will guess the cube root to be 11.1; then we multiply by 10 (since we took at one factor of 1000) to get 111.

How did we do? The answer is 111.5; so we did pretty good! The correct answer range is 106-117.

13.3 Other Estimation Problems (H)

Some approximation problems are multiplication questions that are a matter of rounding off the factors to make it easier:

Example: $59 \cdot 58 \cdot 61 =$

I'm just going to call this $60 \cdot 60 \cdot 60 = 6^3 \cdot 1000 = 216000$

The exact answer is 208742 with the answer range 198305 to 219179.

Another typical problem involves $\pi^2 \approx 9.869588$

Example: Approximate $\pi^2 \cdot 316$

We will use 10 for π^2, yielding $10 \cdot 316 = 3160$

The 5% answer range would be 2963 to 3274, so we are well within that!

Example: Approximate $297 \cdot 304 + 23 = $ _______________

This would be a lot easier if it were $297 \cdot 303$ since we could then use the difference of squares: $(300-3)(300+3) = 90000+9$. We will use that as our answer: 90009. How did we do? The exact answer is 90311 with a range of 85795-94826. So we did quite well!

Chapter 14 Operations with Fractions (EMH)

14.1 Add/Subtracting Using the Cross Product

Adding fractions using the cross product:

$$\frac{a}{b}+\frac{c}{d}=\frac{(ad+bc)}{bd}$$

Trick:

Numerator = cross product of the numerators and denominators

Denominator is product of denominators.

And don't forget to reduce before you write down the answer!

Subtraction follows the same trick, except of course the + sign is changed to a − in the cross product.

Example: $\dfrac{3}{4}+\dfrac{9}{17}=$ ________________

Numerator = 3·17 + 9·4 = 87
Denominator = 4 · 17 = 68

Answer: $\dfrac{87}{68}$

Example: $\dfrac{7}{9} - \dfrac{10}{17} =$ _____________

Numerator= 7·17-9·10= 29

Denominator= 9·17=153

Answer: $\dfrac{29}{153}$

14.2 Comparison Using Cross Product

The cross product is also useful for comparing two fractions to determine which is larger.

Trick: $\dfrac{a}{b} > \dfrac{c}{d}$ if $ad > bc$

Example: Which is larger, $\dfrac{7}{12}$ or $\dfrac{14}{23}$?

Left side: 7·23=161

Right side: 12·14=168

Therefore, the fraction on the right is larger: $\dfrac{14}{23}$

MATH IS FUN!

Chapter 15: Fraction and Mixed Number Tricks (MH)

15.1 Mixed Number Multiplication – whole part the same; fractional part adds to 1

This is a pretty cool trick that makes a difficult looking problem very easy. If you want to work out the algebra, consider this formula:

$$(a+b)(a+(1-b)) = a(a+1) + b(1-b)$$

where a is the whole number part and b and $(1-b)$ are the fractional parts. Note that I arranged b and (1-**b**) to add to 1!

The Trick: The answer will be found in two steps as follows:

Whole number part: Multiply the whole part by itself + 1
Fractional part: Multiply the fractional parts:

Example: $3\dfrac{3}{5} \times 3\dfrac{2}{5} = $ ___________

Solution: Whole = 3 x (3+1) = 12

$$\text{Fraction} = \frac{3}{5} \times \frac{2}{5} = \frac{6}{25}$$

$$\text{Answer: } 12\frac{6}{25}$$

15.2 Eliminating the Fraction

Some multiplication problems beg for you to get rid of the fraction. You can do this by multiplying by 1 in a clever way.

We are making use of the fact that:

$$a \cdot b = a \cdot b \cdot \left(\frac{k}{k}\right),$$ where k is the number we pick to make the problem

easier! Then, put the k's where they help the most (multiply by the fraction, divide into the whole number:

$$ak \cdot \frac{b}{k}$$

Example: $36 \cdot 2\frac{1}{2} =$

If we divide 36 by 2 and multiply $2\frac{1}{2}$ by 2, we have not changed the value of

the expression. This changes the problem to $18 \cdot 5 = 90$, which should be much easier to do!

Example: $364 \cdot 25\frac{1}{4}$

This time, we will divide by 4 and multiply by 4; this changes the expression to:

$91 \cdot 101 = 9191$

That was easy, wasn't it?

15.3 *Shortcut for (a · a/b)*

There are a couple of different cases here depending on whether $\dfrac{a}{b}$ is a proper fraction or not. We are going to rewrite a = b-n, this will make it a proper fraction (assuming n is a positive integer < b).

$$\frac{(b-n)(b-n)}{b} = \frac{b^2 - 2n + n^2}{b} = (b\text{-}2n) + \frac{n_2}{b}$$

So the trick shines forth: The whole number part is going to be (*b*-2*n*); that is subtract 2 times *n* from the denominator; while the fraction will be the difference between the numerator and denominator, squared, as the numerator, while the denominator remains *b*.

Example: $26 \cdot \dfrac{26}{27} = $ ______

The whole number part is 26-2 = 24; numerator is square of 27-26, which equals 1; the denominator is 27. So the answer is: $24 \dfrac{1}{27}$

Example: $21 \cdot \dfrac{21}{31} = $

The whole number part is 31-2·10= 11; numerator is square of (31-21)= 100; denominator is 31; so this becomes $3 \dfrac{7}{31}$, we have to carry the 3 over to the whole part to get the final answer: $14 \dfrac{7}{31}$.

Caution: Note that since the square "overflowed" the denominator, we had to adjust the whole number part. The moral of the story is: compute the fractional part first!

Next case, the one where $\dfrac{a}{b}$ is an improper fraction, uses the same trick except the whole number part is $b+2n$, instead of $b-2n$. Rethink the algebra above a little, and you will see why!

Example: $14 \cdot \dfrac{14}{11} =$

Note that we will obey our caution and figure the fraction first; numerator is square of 14-11; so that's 9; and the denominator is still 11. No carry this time!

Whole number part is $11 + 2 \cdot 3 = 17$; so the final answer is $17 \dfrac{9}{11}$

Chapter 16 Special Fraction Tricks (MH)

The trick with these tricks is going to be recognizing quickly that the trick applies:

Fraction Sequence 1:

$$\frac{1}{n-1} - \frac{1}{n} + \frac{1}{n+1} = \frac{n^2+1}{(n-1)n(n+1)}$$

Example: $\quad \dfrac{1}{5} - \dfrac{1}{6} + \dfrac{1}{7} = \dfrac{6^2+1}{5x6x7} = \dfrac{37}{210}$

Fraction Sequence 2:

$$\frac{1}{n(n+1)} + \frac{1}{(n+1)(n+2)} + \frac{1}{(n+2)(n+3)} = \frac{3}{n(n+3)}$$

Example:

$$\frac{1}{12} + \frac{1}{20} + \frac{1}{30} = \frac{1}{3x4} + \frac{1}{4x5} + \frac{1}{5x6} = \frac{3}{3x6} = \frac{1}{6}$$

Chapter 17 Remainders and Modulo Arithmetic (EMH)

17.1 Remainder Tricks (EMH)

There are easy tricks for quickly determining the remainder when dividing by 3, 4,5,7,9 and 11:

Three (3):

Find the sum of the digits; divide by 3; use remainder

Example: What is the remainder when 751 is divided by 3?

7+5+1=13   13 divided by 3 leaves remainder 1.

Four (4)

Use only the last two digits; divide by 4; use remainder

Example: What is the remainder when 1643 is divided by 4?

Use last two digits (43); 43 divided by 4 leaves remainder 3

Five (5):

Use only last digit; divide by 5; use remainder

Example: What is the remainder when 3256 is divided by 5?

Use last digit (6); 6 divided by 5 leaves remainder 1.

Seven (7):

This is not called for explicitly in any number sense question, but could be useful to you. (I made this trick up).

Example: What is the remainder when 2531 is divided by 7?

Take everything but the last two digits (in this case, 25) multiply by 2 and add the result (50) to the last two digits, yielding 81. Divide this by 7, the remainder is 4.

Nine (9):

Find the sum of all of the digits; divide by 9; use remainder

Example: What is the remainder when 8621 is divided by 9?

8+6+2+1 = 17; 17 divided by 9 leaves remainder 8.

Eleven (11):

Take everything but the last two digits and add to the last two digits. Take the remainder when divided by 11 (Ok, I made this trick up too!)

Example: Find the remainder when 5328 is divided by11.

53 + 28 = 81; remainder when divided by 11 is 4.

Another common remainder problem relies on the following fact:

The remainder of (ab)/n is equal to the remainder of a/n x remainder of b/n.

Example: If a/5 and b/5 each give a remainder of 3, what is the remainder of (ab)/5?

Solution: Multiply the remainders: $3 \cdot 3 = 9$; take the remainder when dividing by 5;

Answer: 4

A somewhat harder problem.

Example: Find the remainder of $5^{23} \div 7$

Since 7 does not divide into 5, it is not going to divide into 5^{23}. There are thus 6 possible remainders, which will repeat in a cycle 5,4,6,2,3,1. From this, we see that we can take out 5^{18}, since 18 is the nearest multiple of 6. This leaves us with computing the remainder of 5^5 divided by 7. Since you have memorized the power tables, you know 5^5 is equal to 3125. Using my divide by 7 trick above, you can determine that the remainder is 3. This one probably took more than 7.5 seconds!

For more information on this topic, try searching for Fermat's little theorem or the Euler Function.

17.2 Modulo Arithmetic (MH)

Modulo arithmetic is straightforward to understand; I put it here with remainders because the two topics are closely related. I will throw in some mathematical symbols here and explain them as well:

Definition: For integers x,a,n and b, $x \bmod a = b \Leftrightarrow \exists n \ni x = an+b$ and $b < a$.

This reads "For integers x,a,n and b, x mod a = b if and only if there exists n such that x=an+b and b < a"

This is really just a hard way of saying that x mod a is the remainder after dividing x into a. Do you get it? Do you wonder why math books are so hard some times?

Examples: 7 mod 3 = 1   4 mod 2 = 0   11 mod 8 = 3

Chapter 18 Common Fractions, Decimal, Percent (EMH)

These common conversions should be memorized!

Fraction	Decimal	Percent
1/16	.0625	6 1/4%
1/15	.0666…	6 2/3 %
1/12	.08333	8 1/3%
1/11	.0909…	9 1/11%
1/10	.1	10%
1/9	.111…	11 1/9%
1/8	.125	12 ½%
1/7		14 2/7%
1/6	.166…	16 2/3%
1/5	.2	20%
¼	.25	25%
1/3	.33….	33 1/3%
½	.50	50%
¾	.75	75%
1/8	.125	12 ½ %

3/8	.375	37 ½%
5/8	.625	62 ½ %
7/8	.875	87 ½%

Chapter 19 Set Theory (MH)

There are a couple of common number sense problems involving sets. One calls for determining how many subsets a set has, the other asks for the number of proper subsets. Both of these are easy to answer once you know the definitions. A few other key definitions are included for completeness.

For our number sense purposes, a *set* can be considered as a collection of numbers. Set theory can get pretty deep (look into Russell's Paradox sometime!) and cause you to doubt everything you know about math! So, let's keep it simple for answering number sense questions.

Definitions:

Let G,H be sets; then G is said to be a *subset* of set H if every element in G is also in H.

G is said to be a *proper subset* of H if G is a subset of H and H contains at least one element not in G.

Every set is an *improper* subset of itself. Therefore, to determine the number of *proper* subsets, subtract 1 from the total number of subsets.

The *empty set* is the set with no elements.

If a set A has n elements, then the number of subsets of A is 2^n. The totality of subsets of a set is sometimes referred to as the *power set*.

Union: A $\bigcup$ B = all elements included in either A or B

Intersection: A $\bigcap$ B = all elements included in both A and B

Complement of X in A = all elements in A that are not in X

The *Cartesian product* of two sets A, B consists of a set of all the ordered pairs formed by taking all possible combinations of elements from A and B. The number of elements in the Cartesian product is the product of the number of elements in sets A and B.

Chapter 20 Difference of Squares (MH)

This chapter might be generally classified as algebra, but it is an important and useful trick that deserves its own treatment.

The factorization of the difference of two perfect squares may be expressed as:

$$a^2 - b^2 = (a-b)(a+b)$$

This formula can be used from either side, so to speak, to simplify a problem:

Example: $47 \cdot 53 = (50 - 3)(50 + 3)$

$$= 50^2 - 3^2$$

$$= 2500 - 9$$

$$= 2491$$

This use of the formula relies on the relative ease of squaring a number that ends in 0, and so would be used for problems like: $89 * 91$, $62 * 58$, etc.

Example: Consider: $56^2 - 44^2$

In this case, neither number is particularly easy to square. But, using the formula from the other direction, you will find the numbers work out pretty nicely!

$$56^2 - 44^2$$

$$= (56 - 44)(56 + 44)$$

$$= 12 \cdot 100$$

$$= 1200$$

Chapter 21 More Squares Tricks (H)

The tricks in this chapter are simplifications based on elementary algebra. If you know much algebra at all, you should be able to prove to yourself why these tricks work!

21.1 $x^2 + y^2 - (x-y)^2$

The trick is to compute 2xy; that will give you the answer! Like I said, a little algebra will show you why.

Example: $30^2 + 16^2 - 14^2 =$

First, ask yourself if the trick applies!? Yes 30-16=14, so it fits the formula.

The answer is thus $2 \cdot 30 \cdot 16 = 960$

That's a lot easier than juggling 3 different squares in your head, isn't it?

21.2 $x^2 - (x-1)^2 + (x-2)^2 + (x-3)^2$ etc.

Another simple algebra trick; here you have alternating positive and negative signs with the number being squared being decreased by one each time. The answer is found by computing 4x-6.

Example: $21^2 - 20^2 + 19^2 - 18^2 =$

Does the trick apply; yes, the signs alternate and the numbers are decreasing consecutive integers. So the answer is:

$4 \cdot 21 - 6 = 78$

We will look at another variant of this type of problem:

Example: $1^2 - 2^2 + 3^2 - 4^2 + 5^2 - 6^2 + 7^2 - 8^2 =$ _______________

This forms an arithmetic series when considered in pairs. You can work through the math on that and derive the formula; assuming the series starts at 1, ends at an even number n; and has the alternating signs:

$$\text{Sum} = \frac{-n}{2}(1+n)$$

$$\text{Sum} = \frac{-8(1+8)}{2} = -36$$

If the number of terms is odd, compute the sum one less (that will be even), then just add in the last term.

21.3 $(x-n)^2 + (x+n)^2$

The sum of two squares that are a given offset n from a number that is easy to square. The answer is found by computing $2 \cdot (x^2 + n^2)$

Example: $27^2 + 33^2 = (30-3)^2 + (30+3)^2$

$$2 \cdot (30^2 + 3^2) = 1818$$

Example: $99^2 + 101^2 = (100-1)^2 + (100+1)^2$

$$2 \cdot (100^2 + 1^2) = 20002$$

21.4 $x^2 + (x+1)^2$

The sum of the squares of two consecutive integers can be found as follows:

x² + (x+1)² = 2x (x+1) + 1

This may give us a good opportunity to use the distributive property!

Example: 94² + 95² =

$$= 2 \cdot 94 \cdot 95 + 1$$
$$= 2 \cdot ((95\text{-}1) \cdot 95) + 1 \quad \text{(why express it this way?)}$$
$$= 2 \cdot (95^2 - 95) + 1 \quad \text{(because it's easy to sqaure 95!)}$$

$$= 2 \cdot (9025\text{-}95) + 1$$

$$= 17861$$

21.4 $x^2 + (x+n)^2$

Generalizing the previous trick to integers that are close together, but not consecutive:

Example: $64^2 + 67^2 =?$

Consider this algebraically:

$$x^2 + (x+n)^2 =$$

$$x^2 + x^2 + 2nx + n^2 =$$

$$2x^2 + 2nx + n^2 =$$
$$2x(x+n) + n^2$$

To put the trick into words:

"To add two squares that are fairly close to each other, multiply the two numbers, double the result, and then add the square of the difference".

Example as above: $2(64)(67) + 9 = 8585$

Example: $27^2 + 31^2 = 2(27)(31) + 16 = 1690$

What if there are 3 squares?

$$x^2 + (x+n)^2 + (x+2n)^2 =$$
$$x^2 + x^2 + 2xn + n^2 + x^2 + 4xn + 4n^2 =$$
$$3x^2 + 6xn + 4n^2 =$$
$$3x(x+2n) + 4n^2$$

To put this trick into words:

"To add 3 squares, with a constant difference, multiply the first and last number, triple the result, and then add 4 times the square of the difference"

Example: $17^2 + 20^2 + 23^2 = 3(17)(23) + 4(9) = 1173 + 36 = 1209$

Chapter 22 Product of 4 Consecutive Integers +1 (H)

This trick is so cool it deserves its own chapter! To use the trick, you are going to have to know your squares

Example: $6 \cdot 7 \cdot 8 \cdot 9 + 1 = $ _______________

Solution: The answer is found by multiplying the first and last numbers, adding 1, and squaring the result. In this case, $(6 \cdot 9 + 1)^2 = 55^2 = 3025$

The general formula is as follows:

$$x \cdot (x+1) \cdot (x+2) \cdot (x+3) + 1 = ((x \cdot (x+3)) + 1)^2$$

Making it Harder: What if you are given the result and asked to find the numbers?

Example: The product of 4 consecutive integers, plus 1, is 1681. What is the smallest of the 4 integers?

Solution: Using the formula backwards, we start by taking the square root of 1681, which is 41. Subtract 1, getting 40. We are now looking for two numbers that differ by 3 and whose product is 40; that would be 5 and 8. Since we are looking for the smallest number, the answer is 5.

Chapter 23 Sequences and Series (MH)

23.1 Arithmetic Series

The story is told of mathematician Karl C.F. Gauss (1777-1855) that he discovered a formula for adding a series of consecutive integers at the age of 6. His teacher had given the class the assignment of finding the sum of the integers from 1 to 100 as busywork. Young Karl allegedly produced the sum in the matter of a few seconds, stunning the teacher and the rest of the students.

You can see how he might have developed the trick as follows:

First, consider the series to be added, then write the same series in

reverse order just under it:

$$1 + \quad 2 + \quad 3 + \, ... \, + 98 + 99 + 100$$

$$100 + 99 + 98 + \, ... \, + \quad 3 + 2 + \quad 1$$

Take a look at the sum of the upper and lower terms in the series. Notice anything special? They each add up to 101! And of course, there are 100 such terms (since we are counting from 1 to 100), so the sum of all of the terms is $100 \cdot 101 = 10,100$. Since we added the series twice, we divide by 2 to get the final answer: $10100/2 = 5050$.

This example yields the following formula:

Sum of the First n Integers:

$$1 + 2 + 3 + ... + n = \underline{n\,(n+1)} \qquad\qquad 2$$

Two other special arithmetic series:

Sum of the First n Odd Integers:

$$1 + 3 + 5 + ... + (2n\text{-}1) = n^2$$

Sum of the First n Even Integers:

$$2 + 4 + 6 + ... + 2n \; = n(n + 1)$$

Example: Find the sum of $1 + 3 + 5 + 7 + 9 + 11 + 13 + 15$

Solution: $2n\text{-}1 = 15$; so $n=8$; and $8^2 = 64$. Therefore, the sum is 64.

Example: Find the sum of $2 + 4 + 6 + 8 + ... + 26 + 28$

Solution: $2n=28$; so $n=14$; $n(n + 1)$ is thus $14 \cdot 15 = 210$

What is an arithmetic series? (Note that 'arithmetic' in this context is pronounced with the accent on the syllable *met*). An arithmetic series is a sequence of numbers in which each term (except for the first term) varies from its predecessor by a constant difference. In our examples above, this constant was 1 (for all of the integers) and 2 (for both the even and odd series).

A general representation of an arithmetic series would thus be:

$$a + (a+d) + (a + 2d) + (a + 3d) + ... (a + (n\text{-}1)d)$$

Here, the difference between each term is d.

The sum of such a general series is given by the formula:

$$S = \underline{n\ (2a + (n-1)d)}$$

The general form of the series can also be used to find a particular term.

Example: Find the 32nd term of the series $2 + 7 + 12 + 17....$

$a_{32} = (a + (n-1)d)$ where a is the first term (in this case 2), n is the number of the term (in this case, 32) and d is the difference (5).

So, $a_{32} = 2 + 31 \cdot 5 = 167$

23.2 Geometric Series

A geometric series is a sequence of numbers in which each term (except for the first term) is a constant multiple of its predecessor.

Example: $1 + 2 + 4 + 8 + 16 + 32 + ... 256$

Here, the constant multiple is 2; the resulting sequence is that of the powers of 2 (beginning with 2^0).

This is a special series that deserves its own formula:

$$1 + 2 + 3 + ... + 2^{(n-1)} = 2^n - 1$$

Thus, in our example, $256 = 2^8$; so the sum is $2^9 - 1 = 511$.

The general form of a geometric series is:

$$a + ar + ar^2 + ar^3 + ... + ar^{(n-1)}$$

To find the sum of the series:

$$S = \frac{a\,(1- r^n)}{1 - r}$$

Note how this simplifies to the special formula above for powers of 2 with a=1 and r=2.

23.3 Infinite Geometric Series

If $|r| < 1$ for an *infinite* geometric series, then the series is said to *converge* and has a sum given by the following formula:

for the series: $a + ar + ar^2 + ar^3 + \ldots + ar^{(n-1)} \ldots$

the sum is given by: $S = \dfrac{a}{1-r}$

Example: Find the sum of $\dfrac{1}{3} + \dfrac{1}{30} + \dfrac{1}{300} + \dfrac{1}{3000}\ldots$

$a = \dfrac{1}{3}$ (the first term); $r = \dfrac{1}{10}$ so $S = \dfrac{\frac{1}{3}}{\frac{9}{10}} = 10/27$

23.4 Finding the Next Number (EMH)

There is no guaranteed way to figure this out, but the following example has been popular on number sense tests for years, so be ready to recognize it!

Example: Find the next number in the sequence: 8,9,13,22,38,x

Note the differences in the terms: 9-8=1; 13-9=4, 22-13=9, 38-22=16 giving us the sequence 1,4, 9,16. Recognize those numbers?   Of course, these are the perfect squares so the next square is 25; we add that to 38 and the answer is 63.

23.5 Sum of Short Series

Consider the following problem:

Example: 47 + 49 + 51 + 53 + 55

Note that this series has the form (51-4) + (51-2) + 51 + (51 + 2) + (51+4)

The —4, 4, -2, and 2 cancel out and you are left with:

$$51 + 51 + 51 + 51 + 51 = 5 \cdot 51 = 255$$

So the trick: Multiply the number of terms by the middle number (assuming there are an odd number of terms).

Chapter 24 Repeating Decimals (MH)

24.1 Introduction

A real number r is said to be *rational* if there exists integers p,q such that

$$r = \frac{p}{q}$$

All such fractional expressions fall into two categories when converted to a decimal:

1) They terminate (for example $\frac{3}{4} = .75$);

2) They form a repeating pattern (for example: $\frac{1}{3} = .333....$)

Any other decimal expression that never forms a repeating pattern is an *irrational* number (two most famous such numbers being π and e). Even though it is hard to point out other example of specific irrational numbers, there are more irrational numbers than rational numbers (this gets into the concept of different infinities, which we will not go into here!).

Your number sense mission is to convert repeating decimals back into fractions. There are two notations for repeating decimals.

1) The series of dots (as above) after the pattern is repeated;
2) or the overline, which explicitly indicates the pattern of digits ($.23\overline{56}$, which indicates that the 56 repeats forever).

24.2 All Digits Repeat

These are the easy ones: if all the digits involved repeat, then the answer is put the pattern in the numerator, and an equal number of 9's in the denominator. Make sure to reduce your fraction before you write it down!

Example: Convert .444... to a fraction.

The pattern is 4 repeats, so we only need one 9, therefore the answer is $\dfrac{4}{9}$

Example: Convert $.\overline{233}$ to a fraction.

The pattern is '233' so we need three 9's, therefore the answer is $\dfrac{233}{999}$

24.3 Not all Digits Repeat

These are a little bit harder! You can see how to derive this trick by using the sum of an infinite geometric series, which is what a repeating fraction is:

$$.2\overline{31} = \frac{2}{10} + \frac{31}{1000} + \frac{31}{100000} \,...$$

Use the formula $S= \dfrac{a}{1-r}$, where a is the first term and r is the ratio, and you can derive this trick. Or, you can take my word for it!

Example: Convert $.2\overline{31}$ to a fraction

The numerator is determined as follows:

Take the non-repeating part (2) and one copy of the repeating part (31); subtract the non-repeating part (2):

231-2= 229

The denominator is a pattern of 9's and 0's: Use the 9's first, and use as many as there are repeating digits (in this case 2); and then use as many 0's as the non-repeating part (in this case 1).

Denominator= 990

So the answer is $\dfrac{229}{999}$

Example: Convert $.23\overline{18}$ to a fraction.

Numerator: 2318-23=2295   Denominator= 9900; this will have to be reduced: $\frac{459}{1980}$. Ok, this one was really hard. Do you get it now?

Chapter 25 Median, Average, Mode (EMH)

25.1 Average

Definition: The *average* (or *mean*) of a set of numbers is the sum of the numbers dividing by the number of numbers.

Example: Find the average of the numbers 16, 21, 13, and 50.

Sum = 16 + 21 + 13 + 50 = 100

Divide by the number of numbers (4): $\dfrac{100}{4} = 25$

25.2 Median

Consider a set of numbers arranged in ascending order. The middle such number is the *median*. If there are an even number of numbers, then the median is the average of the two middle numbers.

Example: Find the median of: 12,8,6, 15, 14, 7, 26

In order: 6,7,812,14,15,26

There are 7 numbers; the middle number being the 4th, so the median is 12.

Example: Find the median of: 50, 28, 32,15,62, 54

In order: 15,28,32,50,54,62

There are 6 numbers; the middle numbers are thus the3rd and 4th (32 and 50); the average of which is 41. Therefore, the median is 41.

25.3 Mode

The *mode* of a set of numbers is the number that appears most frequently.

Example: Find the mode of: 3,7,2,7,3,10,12,7

The number 7 appears 3 times, more than any other number, therefore,

the mode is 7.

Example: Find the average of 84, 73, 85, 79, and 89.

The obvious way to do this is to add up the numbers and divide by 5. But notice that all of the numbers are close to 80. This makes a shortcut practical: add up the differences from 80, divide by 5; and add the result to 80. (Why does this work?)

So we compute: 4-7+5-1+9=10; divide by 5 yields 2; so the average is 82.

Chapter 26 Geometry (MH)

26.1 Formulae and Definitions for Geometry

Square: A = Area   S = length of side P = perimeter   D= diagonal

$$A = s^2 \quad P = 4s \quad D = s \cdot \sqrt{2}$$

Cube: SA = Surface Area V = Volume   S = length of side

$$V = s^3$$

$$SA = 6s^2$$

Rectangle: A = Area l = length w = width P = perimeter

$$A = lw \quad P = 2l + 2w$$

Trapezoid: A = Area b_1, b_2 = bases   h = height

$$A = \tfrac{1}{2}(b_1 + b_2)\, h$$

Rhombus A = area d_1, d_2 = diagonals

$$A = \tfrac{1}{2}(d_1 \cdot d_2)$$

Circle: A = Area   r = radius C = circumference d = diameter

$$A = \pi r^2$$

$$C = 2\pi r = \pi d$$

D = 2r

Triangle: A = area   b= base   h=height

A = ½ bh

Sphere: V = volume SA = surface area r= radius

$$V = \frac{4\pi}{3}\, r^3$$

$$SA = 4\pi r^2$$

Rectangular prism:

V = volume SA = surface Area

l = length w=width h=height

$$V = l \cdot w \cdot h$$

$$SA = 2lw + 2\,l\,h + 2\,hw$$

Right circular cone: V= volume SA= surface area

r= radius (of base circle) h=height

$$V = \frac{\pi}{3}\, r^2 h$$

$$SA =$$

Using the formulas:

Example: The area of a circle is 9π. Find the diameter.

$$A = \pi r^2$$

$$\pi r^2 = 9\pi$$

$$r^2 = 9$$

$$r = 3$$

$$D = 2r = 2 \cdot 3 = 6$$

Example: The perimeter of a square is 24 inches. Find its area.

_______________________________________ in ^{2}

$P = 4s$; so $24 = 4s$; therefore $s = 6$.

$$A = s^2$$

$$A = 36 \text{ in}^2$$

Definitions:

A *right angle* has a measure of $90°$.

Two angles are *supplementary* if the sum of their measures is 180°.

Two angles are *complimentary* if the sum of their measures is 90°.

Example: A, B are complimentary angles; angle A is 54°; find the measure of angle B.
  90° - 54° = 36°

26.2 Polygons

The sum of the exterior angles of any polygon is 360°.

The sum of the interior angles of any polygon is $(n-2) \cdot 180°$, where n = number of sides of the polygon.

Name	Number of Sides	Sum of Interior Angles
Triangle	3	180°
Quadrilateral	4	360°
Pentagon	5	540°
Hexagon	6	720°
Septagon	7	900°
Octagon	8	1080°

Nonagon	9	1260°
Decagon	10	1440°

Example: The ratio of the angles of a triangle is 1:2:7; what is the smallest angle?

From our table, we know that the sum of the angles is 180°. The sum of the given ratios 1+2+7 is 10; so we apportion 180° accordingly to get the angles: 18°, 36°, and 126°. The question calls for the smallest angle, so the answer is 18°.

26.3 Pythagorean Theorem

Theorem: Suppose a right triangle has sides of length a,b,c, where c is the longest side, then

$$a^2 + b^2 = c^2$$

Example: A right triangle has a leg of length 8 and a hypotenuse of length 17, find the length of the other leg.

Solution: $8^2 + b^2 = 17^2$

$$64 + b^2 = 289$$

$$b^2 = 225$$

$$b = 15$$

Common Right Triangles with Integer Lengths: ("Pythagorean Triples")

Leg	Leg	Hypotenuse
3	4	5
5	12	13
6	8	10
7	24	25
8	15	17
9	12	15
12	16	20

14	48	50
20	21	29

Note that if a,b,c is a Pythagorean triple, then so is ka,kb,kc, where k is any positive integer.

Example: 3,4,5 is such a triple; multiply each side by 2 to obtain 6,8,10.

26.4 Geometric Mean

The geometric mean of two numbers is the square root of their product. Thus, the geometric mean of 4 and 9 is $\sqrt{4x9}$ = 6.

Example: Find the geometric mean of 9 and 16.

$(9 \cdot 16) = 144$, take the square root, yielding 12.

26.5 Analytic Geometry

Given two points (x_1, y_1) and (x_2, y_2)

The distance d between the two points is given by:

$$d = \sqrt{[(x_2 - x_1)^2 + (y_2 - y_1)^2}$$

The slope m of a line containing the two points:

$$m = \frac{(y_2 - y_1)}{(x_2 - x_1)} \qquad \text{where } x_2 \neq x_1$$

The midpoint of the segment defined by the two points:

$$\text{Midpoint} = (\frac{(x_1 + x_2)}{2}, \frac{(x_1 + x_2)}{2})$$

The formula for distance between two points should remind you of the Pythagorean Theorem. If you imagine a vertical segment (of length $y_2 - y_1$) connecting the y-coordinates of the points, and a horizontal segment

(of length $x_2 - x_1$) connecting the x-coordinates of the points, then a right triangle is formed. The hypotenuse of this right triangle is the distance between the two points.

Relationships of Lines and slopes:

If m_1 and m_2 are the slopes of two lines,

If $m_1 \cdot m_2 = -1$, then the two lines are perpendicular

If $m_1 = m2$, then the lines are parallel.

A line given by the equation y=mx + b has slope m and y-intercept b and x-intercept $\dfrac{-b}{m}$. The x, y intercepts are the points at which the line crosses the respective axis.

General formula for a circle:

$$(x\text{-}h)^2 + (y\text{-}k)^2 = r^2$$

where the point (h,k) is the center of the circle and r is the length of the radius.

A circle is a special case of the ellipse in which the foci coincide.

An ellipse given by the equation:

$$\frac{(x-h)^2}{a^2} + \frac{(y-k)^2}{b^2} = r^2$$

The area of the ellipse is given by: A= πab

26.6 Polar Coordinates (H)

You are probably familiar with designating points in the plane using two points (x,y) in what is called the *rectangular coordinate* system. This system is also known as the *Cartesian coordinate* system, in honor of French mathematician René Descartes (1596-1650).

This system is based on a point of origin labeled (0,0); all other points are then given by their horizontal (*x* coordinate) and vertical (*y* coordinate) offset from the origin. Thus, the ordered pair (6,8) represents a point that is 6 units horizontally from the origin and 8 units vertically from the origin. If you picture this in your mind, you can visualize a right triangle with one vertex at the origin, another at the point (6,0) and another at (6,8). We know the length of two of these sides (namely, 6 and 8) so we can compute the length of the hypotenuse using the

Pythagorean Theorem. We will refer to this length as r:

$$6^2 + 8^2 = r^2 \quad \text{therefore, r} = 10.$$

Back to visualizing the triangle: the angle at the origin we will refer to as θ, we can find the measure of this angle since we know all of the sides:

$$\tan \theta = \frac{8}{6}$$

$$\text{therefore } \theta = \arctan\left(\frac{4}{3}\right), \text{ so } \theta \approx 59°.$$

What we have just done is to derive the *polar coordinates* (10, 59°) equivalent to the rectangular coordinates (6,8). The polar coordinates represent a radius r taken along the positive x axis, which is then rotated θ degrees.

Let's turn this into formulas:

If (x,y) are rectangular coordinates, then (r, θ) are the corresponding polar coordinates with:

$$r = \sqrt{x^2 + y^2} \quad \text{and} \quad \theta = \arctan\left(\frac{y}{x}\right)$$

Given (r, θ) are polar coordinates of a given point, the rectangular points (x,y) can be found as follows:

$$x = r \cos \theta \quad \text{and} \quad y = r \sin \theta$$

We are now ready to do a couple of number sense problems which will be easy now that you are armed with these definitions:

Example: Change (6, 0°) to rectangular coordinates (x,y).

$y =$ ________

Solution: $y = r \sin \theta$, so $y = 6 \sin 0° = 6 \cdot 0 = 0$

Example: Change (12,9) to polar coordinates (r, θ)

$r =$ _____

Answer: I hope you recognize (12,9) as the two legs of a Pythagorean triple with 15 as the hypotenuse. If you do not, you will have to work through the arithmetic: $r = \sqrt{12^2 + 9^2} = 15$

Note the problem does not call for computing the angle θ, which would have been hard!

Cautionary note: Rectangular coordinates are unique, but polar coordinates are not. Think about it: (r, θ) is going to represent the same point as (r, θ + n· 360°), where n is any positive integer. Consider also the polar coordinates (5, 45°). The same point would be represented by the polar coordinates (5, -315°).

Chapter 27: Trigonometry (H)

27.1 Basic Trig Notation and Values

sin = sine   cos= cosine tan = tangent sec = secant csc = cosecant cot=cotangent

For angles less than 90°, sin θ = cos (90°-θ)

The remainder of the trig functions are related as follows:

$$\tan\theta = \frac{\sin\theta}{\cos\theta} \quad \sec\theta = \frac{1}{\cos\theta} \quad \csc\theta = \frac{1}{\sin\theta} \quad \cot\theta = \frac{1}{\tan\theta}$$

$\sin^2 x$ is spoken as "sine squared of x" and is equivalent to $(\sin x)^2$

arcsin x (spoken "Arc sine of x") is the inverse function of sin x and is also referred to by the notation $\sin^{-1}$ x. These conventions apply to all 6 trig functions.

The basic trigonometric value table; these values should be memorized!

Angle	Radians	Sin	Cos	Tan
0°	0	0	1	0
30°	$\dfrac{\pi}{6}$	$\dfrac{1}{2}$	$\dfrac{\sqrt{3}}{2}$	$\dfrac{\sqrt{3}}{3}$
45°	$\dfrac{\pi}{4}$	$\dfrac{\sqrt{2}}{2}$	$\dfrac{\sqrt{2}}{2}$	1

60°	$\dfrac{\pi}{3}$	$\dfrac{\sqrt{3}}{2}$	$\dfrac{1}{2}$	$\sqrt{3}$
90°	$\dfrac{\pi}{2}$	1	0	∞

Here is the plan for memorizing this table:

1) Learn the principle angles (0°-30°-45°-60°-90°)
2) Learn the values for sine
3) The values of cosine are the same as sine, in the reverse order
4) The values for tan can be derived from sin/cos
5) The values for sec, csc, and cot can be derived from the reciprocals of cos,sin, and tan, respectively.

To learn the values for sin, notice the following pattern:

1) Write down 0,1,2,3,4
2) Take the square root
3) Divide each result by 2
4) This will give you the five values for sin in the table.

27.2 Pythagorean Identities

The primary trigonometric identity:

$$\sin^2\theta + \cos^2\theta = 1$$

The second can be obtained by dividing both sides of the primary identity by $\cos^2\theta$

$$\tan^2\theta + 1 = \sec^2\theta$$

The third can be obtained by dividing both sides of the primary identity by $\sin^2\theta$

$$1 + \cot^2\theta + 1 = \csc^2\theta$$

27.3 Additive Identities

$\sin (x \pm y) = \sin x \cos y \pm \cos x \sin y$

$\cos (x \pm y) = \cos x \cos y \mp \sin x \sin y$

$$\tan (x \pm y) = \frac{\tan x \pm \tan y}{1 \mp \tan x \tan y}$$

The additive identities will often pop up towards the end of a number sense test, usually giving you the right-hand side of the expression.

$$\textbf{Example:} \quad \frac{\tan 31° + \tan 14°}{1 - \tan 31° \tan 14°} = \underline{\hspace{3cm}}$$

At first glance, this equation looks rather daunting, but we recognize it as the expansion of tan (x+y) where x=31° and y=14°, so all we have to do is find the tan 45°. From our deluxe basic value table you memorized a few sections back, you know that tan 45° = 1.

27.4 Double Angle Formulas

These formulas can be derived from the additive identities by setting x=y:

$\sin 2x = 2\sin x \cos x$

$\cos 2x = \cos^2 x - \sin^2 x$

$$\tan 2x = \frac{2 \tan x}{1 - \tan^2 x}$$

27.5 Period

The *period* of the 6 basic trig functions is 2π. Adding any multiple of the period to the angle results in the same functional value, i.e.,

$$\sin x = \sin(x + n2\pi) \text{, where } n \text{ is any integer.}$$

Consider sin (kx); the period is x=2π/k.

Example: Find the period of the graph y = 3 sin (2x)

We set 2x = 2π, therefore x= π, therefore the period is π.

27.6 Sample Problems

Example: The hypotenuse of a 30 60° 90° right triangle is 16. Find the smaller leg.

Solution: The smaller leg is opposite the 30° angle; so:

$$\sin 30° = \frac{x}{16}$$

$$\text{since the } \sin 30° = \frac{1}{2} \text{ , we find that } x=8$$

Example: $\arcsin(1) =$ ___________

Solution: This is just a memory problem; what angle has sin = 1?

The answer is $90°$ or $\dfrac{\pi}{2}$ radians, depending on what units are requested.

Example: If $\cot x = \dfrac{1}{2}$, then tax $x =$ __________

Solution: Since $\cot x$ and $\tan x$ are reciprocal functions, $\tan x = 2$.

Example: $\arctan (\tan 180)) =$ ____________

Solution: This is one of those problems that is so easy, it seems like a trick. The answer is 180 because arctan and tan are inverse functions.

Example: If $\sin 9° = \cos x$, and $0 \leq x \leq 90°$, then x = ____________

Solution: Recall that $\sin x = \cos (90°-x)$, so the answer is $81°$

Chapter 28 Squaring Numbers Near 50 (MH)

This is a homemade sort of trick; just skip it if you do not find it helpful. The more squares you have memorized, the better off you are; this trick may help you quickly recalculate and verify squares you have memorized.

Let u be the difference from 50:

Then $(5u)^2 = ab$

Where a, b are each two digit numbers as follows:

$a = 25 + u$

$b = u^2$ formed as two digits (e.g., $3^2 = $ '09')

Example: $(54)^2$

In this case, u=4, so we find a and b as follows:

a= 25 + 4 = 29

b= 4^2 = 16

Therefore, $(54)^2$ = 2916

Actually, this trick works for any number, but it becomes unwieldy the further you get away from 50!

Consider: $(62)^2$

In this case, u = 12 (the difference from 50), so

a= 25 + 12 = 37

b= 12 ^{2} = 144

Therefore, $(62)^2$ = 3844   (the 1 from 144 being carried over added to the 7)

Also, it works (in reverse) for numbers less than 50:

Example: $(47)^2$

In this case, u = -3 (the difference from 50):

a = 25 – 3 = 22

b = $(-3)^2$ = 09

Thus, $(47)^2$ = 2209.

Chapter 29 Elementary Number Theory (MH)

Number theory refers to the study of the properties of the positive integers and begins with the concept of prime numbers.

Definition: A *prime* number is an integer with exactly 2 positive integral divisors, namely, itself and 1.

The Fundamental Theorem of Arithmetic can be summarized as follows: Each positive integer has a unique factorization as a set of positive prime numbers. This, of course, disregards the order of the factorization. From this basis, we get into the number sense problems.

29.1 Number of Positive Integral Divisors

Definition: Let k, n be integers. Then k is a *positive integral divisor* of n if the remainder of $n \div k = 0$

A common number sense problem asks how many positive integral divisors a particular integer has. The hard way to figure this out for a given number X is to divide all the numbers less than X into X, see which ones give a remainder of 0, and keep track of this count in your head.

Or, you can use combinations, applied to the prime factorization, since each positive integral divisor will be some combination of the prime factors of the number.

Example: How many positive integral divisors does 98 have?

Prime factor $98 = 7^2 \cdot 2$

Now take each exponent from the prime factorization, increase it by 1, and multiply the result.

$(2+1)(1+1) = 6$ (Note that the 2 has an implied exponent of 1.

Therefore, 98 has 6 positive integral divisors, *viz.*, 1,2,7,14,49, 98.

29.2 Sum of Positive Integral Divisors

From our previous example, consider the problem: Find the sum of the positive integral divisors of 98: $1 + 2 + 7 + 14 + 49 + 98 = 171$

Determining all of the factors and then adding them up is going to be rather tedious!

The trick: Determine the prime factorization: $98 = 7^2 \cdot 2$

(Remember that any integer (except 0) to the zero power is 1).

Then, compute the sums $7^2 + 7^1 + 7^0 = 57$

$$2^1 + 2^0 = 3$$

$57 \cdot 3 = 171$ and there is your answer!

29.3 Greatest Common Factor and Least Common Multiple *(MH)*

First the definitions,

Definition: Let k, a, b be integers. Then k is a *common factor* of a and b if the remainder of $a/k = 0$ and the remainder of $b/k=0$. k is the *greatest common factor (GCF)* of a, b if it is a common factor and no other common factor is greater.

(The *greatest common factor* is sometimes referred to as the *greatest common divisor*).

Definition: Let n, a, b be integers. Then n is a *common multiple* of a,b if the remainder of $n/a = 0$ and the remainder of n/b. n is the *least common multiple* *(LCM)* of a,b if it is a common multiple and no other common multiple is less than n.

Example: What is the GCF of 140 and 150?

First, determine the prime factorization of both numbers:

$$140 = 2^2 \cdot 5 \cdot 7 \qquad\qquad 150 = 2 \cdot 3 \cdot 5^2$$

Now, pick each prime factor that appears in both, taking the highest power common:

$$\text{GCF} = 2 \cdot 5 \ = 10$$

Example: What is the LCM of 140 and 150?

Use the prime factorization above; use all of the prime factors, and pick the highest power appearing:

$$2^2 \cdot 3 \cdot 5^2 \cdot 7 = 2100$$

Note that the product of the GCF and LCM of 140 and 150 is 21,000, which is also equal to 140 * 150. This is not a coincidence, and can be stated as a general rule:

$$\text{GCF}(a,b) \cdot \text{LCM}(a,b) = a \cdot b$$

This rule can now be used to easily solve the following types of problems:

Example: What is the product of the GCD and the LCM of 28 and 32?

This is a straightforward application of the formula; $28 \cdot 32 = 896$

(Note that $28 \cdot 32 = (30 -2) \cdot (30 +2)$.

Example: The product of the LCM and GCD of x,y is 375. If y is 25, what is x?

The rule tells us that $25y = 375$; so $y = 15$

Example: The LCM of x, 20 is 300. The GCD of x, 20 is 5. What is x?

The rule tells us that $20x = 1500$; therefore $x = 75$

29.4 The Concept of Relatively Prime

Definition: *a, b* are *relatively prime* if the greatest common factor of *a,b* is 1.

Example: How many integers less than or equal to 20 are relatively prime to 20?

The numbers relatively prime to 20 would be 1,3,7,9,11,13,17, and 19, since the GCF of each of these numbers and 20 is 1.

The shortcut: Obtain the prime factorization: $20 = 2^2 \cdot 5$

Decrease one of each factor by 1 $2 \cdot 1 \cdot 4 = 8$

Example: How many integers less than are equal to 36 are relatively prime to 36?

$36 = 2^2 \cdot 3^2$

Applying the trick: $(2-1) \cdot 2 \cdot (3-1) \cdot 3 = 12$

29.5 The Even Prime

There is one even prime number, namely 2. Believe it or not, this was at one time a number sense question!

Example: Write down an even prime: ___________

Answer: 2

29.6 Triangular numbers, including sums thereof!

The *triangular numbers* are given by the following sequence:

$$1,3,6,10,15, .. \ T_n$$

$$\text{where } T_n = \frac{n(n+1)}{2}$$

The term *triangular* comes from the right triangle pattern that can be formed from the corresponding number of objects.

Example: Find the 7th Triangular number

$$T_7 = \frac{7(7+1)}{2} = 28$$

An interesting fact about the sum of two consecutive triangular numbers:

$$T_{n-1} + T_n = \frac{(n-1)n}{2} + \frac{n(n+1)}{2} = \frac{n^2 - n + n^2 + n}{2} = n^2$$

Example: Find the sum of the 11th and 12th triangular numbers.

Sounds hard, but now that we know the trick, it's easy!

$$T_{11} + T_{12} = 12^2 = 144$$

29.7 Pentagonal Numbers

The *pentagonal numbers* are given by the following sequence:

$$1,5,12,22,... P_n$$

$$\text{where } P_n = \frac{n(3n-1)}{2}$$

Example: Find the 5th pentagonal number.

$$P_5 = \frac{5(3x5-1)}{2} = 35$$

The term *pentagonal* comes from the regular pentagonal pattern that can be formed from the corresponding number of objects.

29.8 Octagonal Numbers

The *octagonal numbers* are given by the following sequence:

$$1, 8,21,40,65,96,133, 176, O_n$$

$$\text{where } O_n = 3n^2 - 2n$$

The term *octagonal* comes from the regular octagonal pattern that can be formed from the corresponding number of objects.

You are probably detecting a pattern here. Yes, there is a general formula for all such polygonal numbers!

Chapter 30 Putting Radicals In Their Place! (H)

We start with the notational convention:

$$\sqrt[n]{a^k} = a^{\frac{k}{n}}$$

The expression on the left hand side is read "The nth root of a to the kth power".

The symbol "$\sqrt{}$" is referred to as a *radical*; the number n is referred to as the *index*. If there is no index, the implied index is 2.

30.1 Basic Principles

There are two basic conventions for radicals on which most number sense problems are based

1) No radicals are left in the denominator of a fraction;
2) No reducible value is left inside a radical

We will give a simple example of each.

Example: $\dfrac{2}{\sqrt{3}} = \underline{\hspace{3cm}}$

Here we have a radical in the denominator, which means the expression is not in simplest form. To eliminate the radical, we multiply by 1 in the form of $\dfrac{\sqrt{3}}{\sqrt{3}}$

$$\frac{2}{\sqrt{3}} \cdot \frac{\sqrt{3}}{\sqrt{3}} = \frac{2\sqrt{3}}{3}$$

Example: $\sqrt[3]{162} =$ _________

Here, there is a perfect cube within the radical that must be extracted:

$$\sqrt[3]{6x27} = \sqrt[3]{27} \cdot \sqrt[3]{6} = 3\sqrt[3]{6}$$

30.2 Difference of Squares with Radicals

Remember the trick for differences of squares? It makes the following problem easy, even though it looks a little tricky at first:

Example: $(\sqrt{7} + \sqrt{3})(\sqrt{7} - \sqrt{3})=$ _______________________

We recall that $(a-b)(a+b) = a^2 + b^2$

So the answer is 7-3=4. That was really pretty easy, wasn't it?

Now see if you can reduce this expression to proper form:

Example: $\dfrac{\sqrt{32}}{3 - \sqrt{2}} =$ ___________________

$$\frac{\sqrt{32}}{3 - \sqrt{2}} \cdot \frac{3 + \sqrt{2}}{3 + \sqrt{2}} =$$

$$\frac{\sqrt{32}(3 + \sqrt{2})}{(3 - \sqrt{2})(3 + \sqrt{2})} =$$

$$\frac{\sqrt{16x2}(3 + \sqrt{2})}{9 - 2} =$$

$$\frac{\sqrt{16}\sqrt{2}(3 + \sqrt{2})}{7} =$$

$$\frac{4(3\sqrt{2} + 2)}{7}$$

Chapter 31 Complex Numbers (H)

You probably spent most of your earlier mathematical years hearing that "there is no such thing as the square root of a negative number."

Well, that's not true. If you make the following definition:

$$\sqrt{-1} = i$$

you can do a lot of math from there! A multiple of i is called an *imaginary* number, but this does not mean it is not real in the usual sense. A number after all, is an idea and this square root of negative 1 thing is a neat idea! Imaginary numbers are not real numbers in the formal sense of real numbers (the numbers represented on a standard number line).

A *complex number* has a *real* part and an *imaginary* part:

with the general form: a + bi

(a is the real part; bi is the imaginary part)

The powers of i repeat in a cycle of 4 numbers:

$$i^1 = i$$

$$i^2 = -1$$

$$i^3 = -i$$

$$i^4 = 1$$

Thus, to compute any power of i, take the remainder of the exponent after dividing by 4 and use the table above:

$$i^{34} = i^2 = -1$$

Addition of complex numbers: $(a + bi) + (c + di) = (a+c) + (b+d)i$

Multiplication of complex numbers $(a + bi)\,(c + di) = (ac - bd) + (ad + bc)i$

The *conjugate* of a complex number $(a+bi)$ is $(a-bi)$.

Note the product of a number and its conjugate has no imaginary part:

$$(a+bi)\,(a - bi) = a^2 + b^2$$

The conjugate is useful for simplifying expression as follows:

Example: Write $\dfrac{3 + 2i}{4 - 3i}$ in terms of a+bi.

Solution: Multiply by 1 in the form of the conjugate divided by itself:

$$\frac{3 + 2i}{4 - 3i} \cdot \frac{4 + 3i}{4 + 3i} = \frac{12 + 9i + 8i - 6}{25} = \frac{6 + 17i}{25} = \frac{6}{25} + \frac{17i}{25}$$

Chapter 32: Matrix Theory

Matrix theory takes the basic linear equations you studied in algebra, puts the coefficients into a box called a matrix; and then things get really complicated!

We will provide here the method for computing the determinant of a matrix and let it go at that.

The *determinant* of a matrix is computed as follows:

$$\begin{vmatrix} a & c \\ b & d \end{vmatrix} = ad - bc$$

Example:

Find the determinant:

$$\begin{vmatrix} -1 & -3 \\ 3 & 2 \end{vmatrix} = -2 - (-9) = 7$$

Chapter 33 Basic Algebra (MH)

The essence of algebra is to read a problem, set up an equation corresponding to the problem, and solve the equation.

33.1 Some Basics

In this section, we will cover a few examples of basic algebra problems.

Example: The product of two numbers is 40, the sum is 13. What is the smaller number?

Algrebraic Solution: This one is so easy you can do it in your head! You can write the equations with one variable by letting x and 13- x represent the two numbers (since the add to 13). Using these in the product, we get:

$$x(13 - x) = 40$$

Which yields the quadratic equation:

$$x^2 - 13x + 40 = 0$$
$$(x - 5)(x - 8) = 0$$
$$x = 5$$
$$x = 8$$

Therefore, the smaller number is 5. So, this would be a lot of work using algebra. For this type of problems, you are better off trying to factor 40 and see which ones add up to 13!

Classic number sense problem: If A can mow the lawn in 5 hours and B can mow the lawn is 8 hours, how long will it take if they work together?

Solution: Of course, there are certain assumptions built into this

problem, but we will gloss over those. Assume the answer is x, then A will do $\dfrac{x}{5}$

of the work and $\dfrac{x}{8}$ of the work, and together they will do 100% of the work, represented by 1, and yielding the equation:

$$\frac{x}{5} + \frac{x}{8} = 1$$

Solving for x yields $\dfrac{40}{13}$ or $3\dfrac{1}{13}$. Note that this is the product of the two numbers (8 and 5) divided by the sum. This turns out to be the general formula for this type of problem.

For completeness, we include two more basic formulas:

Distance = Rate · Time

Interest = Principal · Rate · Period

33.2 Ratio and proportion

These problems are usually a matter of setting up simple equations. As always, make sure you answer the question!

A classic series of number sense problems use the term "y varies... with x", each of which yields an equation as indicated in the table below:

Sentence Form	Equation
y varies directly with x	y = kx

y varies inversely with x	$y = \dfrac{k}{x}$
y varies jointly with x and z	$y = kxz$
y varies directly with the square of x	$y = kx^2$

Example: If y varies directly with x, and y =4 when x=2, find y when x=6.

Solution: We use the first set of data to find k:

y=kx; so 4=k·2; therefore k=2.

So we use the value of k and x to find y:

y=kx; so y=2·6=12

A typical ratio problem:

Example: If 12 apples cost \$1.80, then 14 apples would cost _______ cents.

The equation for this problem would be: $\dfrac{12}{14} = \dfrac{\$1.80}{x}$. We could solve both sides for x, but it would easier in this case to just divide 12 into \$1.80 to get the cost per apple (\$0.15) and then multiply that by 14 to yield \$2.10. Of course, if you write that down, it's wrong because the answer calls for cents! Therefore, we would write 210.

33.3 Evaluating a function

Evaluating a function is simple!

Example: Evaluate $f(2)$ if $f(x) = x^2 - 3x$

We plug in 2 for x:

$$f(2) \quad = 2^2 - 3(2) = 4 - 6 = -2$$

Some problems call for you to evaluate 2 functions in a row:

Example: Evaluate $g(f(8))$ if $g(x) = x^2$ and $f(x) = 3x - 7$

First, we find $f(8) = 3(8) - 7 = 17$. We substitute this value into g:

Now we find $g(17) = 17^2 = 289$

33.4 Inequalities

Inequalities are solved much like equations with one additional rule: if the inequality is multiplied by a negative number, the sense of the inequality reverses.

By *reversing the sense* we mean changing "greater than" to "less than" or vice versa.

Example: If $-2x - 6 > 0$, then x < ____________

$-2x > 6$

Divide both sides by -2 will change the > to <

x < -3

33.5 Quadratic equations/formula

The term *quadratic* refers to polynomials or equations having a square term as the highest power.

The standard form of a quadratic equation is:

$$ax^2 + bx + c = 0$$

The quadratic formula gives the two roots of any quadratic equation:

$$x = \frac{-b \pm \sqrt{b^2 - 4ac}}{2a}$$

$b^2 - 4ac$ is referred to as the *discriminant*.

$b^2 - 4ac > 0$ then the equation has 2 real roots

$b^2 - 4ac = 0$ then the equation has 2 identical roots

$b^2 - 4ac < 0$ then the equation has 2 imaginary roots

Example: Find the discriminant of $x^2 - 3x + 4 = 0$

$b^2 - 4ac = (-3)^2 - 4(1)(4) = 9 - 16 = -7$

33.6 The sum and product of roots for quadratic equations

For a quadratic equation in standard form, the sum and product of the roots is easily determined from the coefficients:

$$\text{Sum of roots} = \frac{-b}{a} \qquad \text{Product of roots:} \ \frac{c}{a}$$

Example: Find the sum and product of the roots for the quadratic equation:

$$3x^2 - 2x - 6 = 0$$

$$\text{Sum} = \frac{-(-2)}{3} = \frac{2}{3}$$

$$\text{Product} = \frac{-6}{3} = -2$$

33.7 Product of roots for cubic equation

The standard form of a cubic equation is:

$$ax^3 + bx^2 + cx + d = 0$$

The general formula for solving cubic equations is pretty complicated and you will not encounter any need for it on a number sense test. However, you may be asked to find the product of the roots which is easy enough:

$$\text{Product of roots:} \quad \frac{d}{a}$$

33.8 Inverse functions

A function $f^{-1}(x)$ is the inverse of a function $f(x)$ if $f(f^{-1}(x)) = x$ for all values of x within the domain of f.

Example of an inverse function for a simple linear equation:

$$\text{If } f(x) = ax + b, \text{ then } f^{-1}(x) = \frac{x - b}{a}$$

Example: Find the inverse function of $f(x) = 2x + 3$

$$f^{-1}(x) = \frac{x - 3}{2}$$

Do a sample test of the relationship $f^{-1}(f(x))$ with $x = 7$

Note that $f(x) = 17$; and then $f^{-1}(17) = 7$ (back to the original value of x)

33.9 Polynomial Remainders

Dividing polynomials into other polynomials is not much fun, fortunately, you do not have to do that to find the remainder when dividing a linear expression like *(x-k)* into a polynomial. It is this simple: The remainder when $f(x)$ is divided by *(x-k)* is $f(k)$. Note that if $f(k)=0$, then k is a root of the polynomial.

Example: Given the polynomial $x^3 + 6x^2 + 4x - 8$; what is remainder when divided by *x+2*?

Solution: The only tricky to part to watch for is to remember that the rule said x-k; here we have *x+2*, so that is *x-* (-2); so our *k* is –2, not 2.

Answer: $(-2)^3 + 6(-2)^2 + 4(-2) - 8 = -8 + 24 - 8 - 8 = 0$

What do you know, -2 is a root of the polynomial!

Chapter 34 Exponents and Logarithms

34.1 Laws of Exponents

We take up exponents and logarithms in the same chapter because they are inverse functions.

Exponents are just a short hand for multiplication:

$$a^4 = a \cdot a \cdot a \cdot a \quad (a \text{ is the base, 4 is the exponent, or power})$$

If you keep this in mind, all of the Laws of Exponents make sense!

Multiplying Base Law: $a^n \cdot a^m = a^{(m+n)}$

Division Law: $\dfrac{a^n}{a^m} = a^{(n-m)}$

Multiplying Exponents Law: $(a^n)^m = a^{(m \cdot n)}$

Negative exponent Law: $a^n = \dfrac{1}{a^n}$

Fractional Exponent Law: $a^{\frac{m}{n}} = \sqrt[n]{a^m}$

Zero Exponent Law: $a^0 = 1$, with a not equal to 0. 0^0 is not defined.

Thought exercise: Use the Division Law with n=m and think about why $a^0 = 1$. Got it?

Caution: Don't make up your own laws of exponents!

Bad Example: $3^n \cdot 4^m = ?$

Well, there is no law to cover this. The bases are not the same! Don't think that it is $12^{(m+n)}$ or anything like that!

Example: If $16^x = 4$, then $x^4 = $ ________

Here, we need to solve for x. First, we need to get both sides in a common base so we can apply a law of exponents:

Since $16 = 4^2$, we can rewrite the problem as:

$4^{2x} = 4$, the exponent of the 4 on the right hand side is an implied 1, so

$2x=1$ or $x=\dfrac{1}{2}$. We now substitute for x in x^4 to get $\dfrac{1}{16}$.

Example: $\dfrac{8^4}{4^3} =$

Remember our caution above! Before we apply the laws, we have to get the base common, in this case 2 looks like a good candidate!

$$\frac{8^4}{4^3} = \frac{(2^3)^4}{(2^2)^3} = \text{ first, we work on our base.}$$

$$\frac{2^{12}}{2^6} = \text{ we use the law of multiplying exponents}$$

$$2^6 = \text{ we use the law of division}$$

$$64$$

Example: $3^{(x+1)} = 63$, then what is 3^x ? ———

Ok, this looks tricky at first, but all we have to do is divide both sides by 3. This subtracts one from the exponent on the left hand side, leaving 3^x.

On the right, we have $63/3=21$.

Therefore, $3^x = 21$.

Example: $2^n - 7 = 1$, then $2 - 7^n =$ _________

This is one of those questions that is visually tricky, but it is actually pretty easy. We are going to solve for n in the first expression, then use it to evaluate the second expression:

$$2^n - 7 = 1$$

$$2^n = 8$$

$$n = 3$$

So, $2 - 7^3 = 2 - 343 = -341$

34.2 Logarithms – Definition and Notation

As discussed earlier, the logarithm function is the inverse of the corresponding exponent function.

Definition: $\log_a b = n$ if $a^n = b$

The a in both expressions is referred to as the *base*. In words, you would read this definition as "The log of b base a is n if a to the nth power is equal to b".

A few words about notation are in order. The number e, which is defined in our chapter on calculus, is the base for what are referred to as *natural* logarithms. Mathematicians have routinely used the notation "log b" to refer to such logarithms with e being the implied base. However, in most ordinary contexts, including number sense problems, the notation "log b" is used when the implicit base is 10, while "ln b" is used to indicate the natural logarithm. So, we will use "log b" for base 10 and "ln b" for natural logarithm throughout the rest of this treatise.

Read the definition of the logarithm again. Do you see that logarithm and exponentiation are inverse functions?

Consider this problem, once a common "trick" question on number sense tests:

Example: $2^{\log_2 3} =$ _______________

If you read this problem, the answer should be obvious: "2 raised to the power that is defined as the power to which 2 is raised to obtain 3". Do you see why the answer is 3?

34.3 Properties of Logarithms – Why the Slide Rule Died

The properties of logarithms are related to the laws of exponents:

Multiplication: $\log_k a + \log_k b = \log_k (ab)$

(Note the base k is present in each term)

Exponents: $\log_k a^n = n \cdot \log_k a$

Division: $\log_k a - \log_k b = \log_k \left(\dfrac{a}{b}\right)$

These properties were at one time very useful for doing calculations. This formed the basis for the slide rule (*requiescat in pace*), a common tool for computation before the advent of the electronic calculator back in the 1970's.

The following table illustrates the exponent property:

Illustrative Log Table #1 – Base 10

The $\log_{10}$ of....	is equal to...	because ..
1	0	$10^0 = 1$
10	1	$10^1 = 10$
100	2	$10^2 = 100$
1000	3	$10^3 = 1000$
10000	4	$10^4 = 10000$

Illustrative Log Table #2   - base 2

The $\log_2$ of....	is equal to...	because ..

1	0	$2^0 = 1$
2	1	$2^1 = 2$
4	2	$2^2 = 4$
8	3	$2^3 = 8$
16	4	$2^4 = 16$

Now for some typical number sense problems involving logarithms:

Example: $\log 1000 = $ ___________

Remember, no base implies base 10. Referring to our log table #1 above, we see the answer is 3.

Example: $\log_3 27 = $ _________

Ok, so this asks the question "to what power is 3 raised to get 27?". That would be 3!

Example: $\log_x 25 = 2$

This problem asks the question, "if $x^2 = 25$, then x=?". So, the answer is 5.

Example: $\ln e^{11} = $ _________

Using our logarithm properties, we convert the expression to:

$$11 \cdot \ln e = 11 \text{ since } \ln e = 1$$

Example: $\log 40 + \log 25 = $ _______

The base is implied, so we assume base 10. Using our multiplication property, we get:

log 40 + log 25= log 1000 = 3

Example: $\dfrac{\log_3 27}{\log_3 9}$

Don't fall for any tricks; you can't just divide 9 into 27 or cancel the logs! Also, our rule for division does not apply, that relates to a difference of logarithms.

Here, we just compute each log separately and write down the resulting fraction:

$\dfrac{3}{2}$

Exercises:

1. $\log 2000 - \log 20 = $ __________
2. $\log_4 x = 4$, then $x = $ __________
3. If $5^{(x+2)} = 775$, then $5^x = $ __________
4. $(\log_2 2^{11})(\log_3 3^{72}) = $ __________
5. $\log_x 6561 = 4$, $x = $ __________

Answers:

1. 2
2. 256
3. 31
4. 792
5. 9

Chapter 35 Probability and Odds (EMH)

What are the odds you will ever use anything from this chapter? Actually, on any given number sense test, the chances are pretty good that you will confront some type of probability question, even at the elementary level.

Probability problems are problems involving fractions in which you must determine the numerator and the denominator.

35.1 Probability

The *probability* P of an event can be expressed: $P = S/T$

where S = successes and T = total possible outcomes. This assumes that each possible outcome is *equiprobable*, i.e., any given outcome is as likely as any other. In actual problems this may be indicated or emphasized by use of terms like "fair coin" or "standard card deck".

Example: A coin is tossed; what is the probability of it coming up "heads"?

Here, $T=2$, since there are two possible outcomes, heads or tails. We have defined one of the outcomes (heads) as success, so $S=1$.

Therefore $P= \frac{1}{2}$.

Other coin problems typically involve successive trials, which bring in the concept of *independent* events. Coin flips are a typical example of independent events; no matter how many times the coin comes up "heads", this does not decrease the likelihood of the same outcome on the next flip. To obtain the probability of a particular set of outcomes from such successive trials, find the product of the individual probabilities.

Example: A coin is flipped 6 times; what is the probability of it coming up "heads" each time?

We already know the probability of each event is $\frac{1}{2}$. The overall probability is therefore: $P = (1/2)^6 = 1/64$.

Card deck problems: A standard card deck has 52 cards in four "suits" (groups) of cards: Clubs, Diamonds, Hearts, and Spades, each suit having 13 cards. The Diamonds and Hearts are red while the Clubs and Spades are black. Each suit has 10 cards numbered one through 10 (although the "one" card is always referred to as the "ace") and 3 face cards: the Jack, Queen, and King.

Example: What is the probability of drawing a red card from a standard card deck?

S = number of red cards = 26 (Hearts + Diamonds); T = total number of cards (52)

Therefore $P = 26/52 = \frac{1}{2}$

Example: What is the probability of drawing an ace from a standard card deck?

S = number of aces = 4 (each suit has one ace); T = total number of cards = 52

Therefore $P = 4/52 = 1/13$.

Besides the coin and standard card deck, you may be confronted with a few other probability problems, typically involving drawing marbles out of a bag. Here, there is no standard set of successes and total possible outcomes as with a card deck, so the problem statement must define those for you.

Example: What is the probability of drawing a blue marble from a black bag that contains 12 blue and 13 red marbles?

Your first question should be: Hey, what does the color of the bag have to do with this? Well, nothing unless it was intended to make it clear that the bag is not clear and therefore the drawings are "fair".

But the problem itself is easy: $S = 12$ since there are 12 blue marbles; $T = 12 + 13 = 25$ total marbles.

Thus $P = 12/25$

Replacement: Probability problems involving successive trials may use the term 'with replacement' or 'without replacement'. This refers to whether the item drawn or selected in a trial is removed from consideration for the next trial.

Example: What is the probability of drawing two blue marbles, without replacement, from a black bag that contains 10 blue marbles and 10 red marbles?

The probability of the first event is computed as follows:

$S = 10$ (blue marbles) $T = 20$ (total number of marbles)

So $P_1 = 10/20 = \frac{1}{2}$

For the second drawing, there is one less marble:

$S = 9$ (blue marbles remaining) $T = 19$ (total number of marbles)

So $P_2 = 9/19$

The overall probability is therefore:

$$P_1 \cdot P_2 = \frac{1}{2} \cdot 9/19 = 9/38$$

If the problem were restated to be 'with replacement', then both P_1 and P_2 would be $\frac{1}{2}$, so the overall probability would be $\frac{1}{4}$.

Another type of problem may ask for the total possible outcomes instead of the probability.

Example: If three coins are tossed, what is the total number of possible outcomes?

Each coin has two possible outcomes; therefore, the total number of possible outcomes is $2 \cdot 2 \cdot 2 = 8$.

Odds: One last topic: occasionally the problem calls for the "odds" to be given rather than the probability.

Recalling our formula for probability: $P = S/T$

The odds is stated as a ratio

$$\text{Odds} = S : (T\text{-}S)$$

Example: If $P = \frac{1}{4}$, the odds are computed as follows:

$S = 1$ $T\text{-}S = 4\text{-}1 = 3$, therefore the odds are 1:3 or "1 to 3".

35.2 Table of Probabilities for Two Dice (MH)

A common number sense problem asked for the probability of rolling a particular number with two fair dice. This problem is so common you are better off just memorizing the table of probabilities. The denominator in all cases is 36 (since there are 6 possible outcomes for each die; $6 \cdot 6 = 36$ gives the total possible outcomes)

Dice Total	Number of Occurrences	Reduced Fraction
2	1	$\dfrac{1}{36}$
3	2	$\dfrac{1}{18}$
4	3	$\dfrac{1}{12}$
5	4	$\dfrac{1}{9}$
6	5	$\dfrac{5}{36}$
7	6	$\dfrac{1}{6}$
8	5	$\dfrac{1}{9}$
9	4	$\dfrac{1}{9}$

10	3	$\dfrac{1}{12}$
11	2	$\dfrac{1}{18}$
12	1	$\dfrac{1}{36}$

This table is fairly easy to memorize if you see the pattern. For the dice totals 2-7, the number of successful outcomes is 1 less than the dice total. You then divide by 36 (being careful to reduce!) and you have the probability for that number. The numbers 7-12 are the same as 2-7 in reverse order.

Note that the singular form of dice is "die".

Chapter 36 Counting Problems – Combinations and Permutations (H)

36.1 Factorials – It's Exciting! (H)

We start with the definition of factorial and use a cool recursive formula as follows:

Definition: $n! = n \cdot (n\text{-}1)!$ for $n > 1$; and $1! = 1$; also $0!=1$.

The term $n!$ is articulated as "n factorial"; the exclamation does NOT mean to yell the letter 'n' in an excited voice!

A more straightforward definition of factorial:

$$n! = n \cdot (n\text{-}1) \cdot (n\text{-}2) \ldots \cdot 3 \cdot 2 \cdot 1$$

Factorials are involved in several further concepts used as number sense problems.

Example: If $7! = 42 \cdot n!$, what is n?

$7! = 7 \cdot 6 \cdot n!$, then $n=5$.

Example: $5! =$

$5 \cdot 4 \cdot 3 \cdot 2 \cdot 1 = 120$

36.2 Combinations: "n Choose k" problems

Definition: A *combination* is an unordered arrangement of a set of objects.

Counting problems come up in a variety of ways in number sense. The most straightforward calls for evaluation of the formula for determining the number of combinations of k items selected from a set of n items (where $k \leq n$):

$$_nC_k = \frac{n!}{k!(n-k)!}$$

$_nC_k =$ is articulated as "n choose k".

Example: $_7C_5 =$

$$= \frac{7!}{5!(7-5)!}$$

$$= \frac{7 x 6 x 5!}{5! x 2!}$$

$$= \frac{7 x 6}{2}$$

$$= 21$$

Example: How many different ways can 6 people be seated in 3 chairs?

$$_6C_3 =$$

$$= \frac{6!}{3!(6-3)!}$$

$$= \frac{6x5x4x3!}{3!x3!}$$

$$= \frac{6x5x4}{3x2}$$

$$= 20$$

36.3 Permutations

Contrast this definition with that of "combinations" above:

Definition: A *permutation* is an ordered arrangement of a set of objects.

Example of a permutation:

How many 4-digit numbers can be made from the digits 1,2,3,4 without repeating any digits?

Well, a number is inherently order dependent, right? So this is a permutation problem, not a combination problem.

Thinking through it intuitively, there will be 4 digits from which to pick for the first slot, 3 remaining for the 2nd slot, 2 for the 3rd slot, and then 1 digit for the final slot.

This gives 4 x 3 x 2 x 1 = 24 permutations. From this example, it is easy to see that the formula for computing the number of permutations of a fixed set of objects is:

$$_nP_n = n!$$

This is a simplification of the more general formula for finding the number of permutations for a subset:

$$_nP_k = \frac{n!}{(n-k)!}$$

 Summary: When order does not matter, it's a combination; when order does matter, it's a permutation.

One other twist that may pop up on a number sense test: how many ways can objects be placed in a circle? This eliminates some of the permutations (think about that for a second and you will see why).

Example: How many ways can 7 objects be placed in a circle?

 Answer: 6! = 720

Chapter 37 Binomial Expansion

There are two typical number sense problems that arise from binomial expansion; one asks you to find a specific coefficient and the other asks for the sum of all the coefficients.

Example: Find the coefficient of x^3y^2 in the binomial expansion of $(x+y)^5$

A general way to express the binomial $(x+y)^n$:

$$(x + y)^n = \sum_{k=0}^{n} {}_nC_k\left(x^k y^{n-k}\right)$$

We can use this formula then to find the specific coefficient we are looking for:

$$_5C_3 = \frac{5!}{(5-3)!3!} = 10$$

Example: Find the sum of the coefficients of the binomial expansion of $(x+y)^6$

The general formula for finding the sum of the coefficients:

$$\sum_{k=0}^{n} {}_nC_k = 2^n$$

The sum of the coefficients for our example is thus:

$$2^6 = 64$$

Back to our first example, what if there are coefficients (other than the implied 1) in front of x and y?

Example: Find the coefficient of x^4y^3 in the binomial expansion of $(2x +3y)^7$

First, determine the binomial coefficient: $_7C_4 = \frac{7!}{(7-4)!4!} = 35$

We must multiply this result by 2^4 and 3^3 (do you see why?)

So the final answer is $35 \cdot 16 \cdot 27 = 15120$

Hopefully, they do not give you a problem quite that hard!

Example: Find the coefficient of the 7th term of $(x+y)^9$ The seventh term will be x^3y^6

So our formula is: $_9C_3 = \dfrac{9!}{(9-3)!6!} = 84$

Chapter 38: Calculus - It Rocks! (H)

The word 'calculus' comes from the Latin word for 'stone' since stones were used as early counting devices. Rather than saying it rocks, most students would probably groan that calculus is hard (as a rock!).

The calculus problems that appear on number sense tests are not that hard and few basic can get you some easy points. I learned these concepts while practicing number sense before I ever had a calculus course.

38.1 Limits

The concept of limits can get complicated, but most number sense limits problems are really just factoring problems.

Example:

$$\lim \frac{x^2 - 3x - 2}{x - 1} =$$

$$x \to 1$$

This reads "The limit of f(x) as x approaches 1"

Note that if the polynomial is evaluated with x=1 as is, you will get a 0 in the denominator, which is not going to be helpful. So instead, we factor the polynomial and reduce it to lowest terms:

$$\lim \frac{(x - 2)(x - 1)}{x - 1} =$$

$$x \to 1$$

$$\lim \ (x-2)$$

$$x \to 1$$

Now, to evaluate the limit, we simply substitute 1 for x to get:

1-2 = -1.

Example: Evaluate: $\lim \dfrac{(x^2 - 4)}{x - 2} =$

$x \to 2$

$$\lim x + 2 = 4$$

$x \to 2$

Definition of the base for natural logarithms:

The number e is the base for natural logarithm and is most commonly defined using limits as follows:

$$\lim (1 + \frac{1}{n})^n = e$$

$n \to \infty$

e is an irrational number whose decimal expression begins: 2.71828.

38.2 Vertical Asymptotes

The limit problems in our previous examples were simple once we did the factoring. Vertical asymptotes bring in another level of complication, along with the concept of infinity. (Note: ∞ is the symbol for infinity).

Definition. Let f be a function which is defined on some open interval containing k except possibly at x = k. We write

$$\lim_{x \to k} f(x) = \infty$$

if f(x) grows arbitrarily large by choosing x sufficiently close to k.

The line x=k is then said to be a *vertical asymptote* of f(x)

Example: Find the vertical asymptote of $\dfrac{-8x+9}{2x-6}$

Difficult concept, but easy to find the answer. We have here a polynomial fraction that cannot be reduced. Where will the value of the function get very large? When 2x-6 approaches 0, since the smaller the denominator, the larger the fraction. So, 2x-6 = 0 fits our definition, i.e., where x=3. Therefore, x=3 is a vertical asymptote of the function.

Example: Find the vertical asymptote of $\dfrac{x^2+4x+3}{3x-11}$

Let's summarize this: For most number sense problems, finding the vertical asymptote is going to be a matter of setting the expression in denominator equal to 0, then solving for x:

$$3x - 11 = 0 \text{ so the vertical asymptote is } x = \frac{11}{3}$$

38.3 Derivative of Polynomials

Consider the polynomial:

$$y = f(x) = k_n x^n + k_{(n-1)} x^{(n-1)} + \dots k_1 x + k_0$$

The derivative of y, referred to as either $\dfrac{dy}{dx}$ or f'(x) is computed by multiplying each exponent of x by the corresponding coefficient and then decreasing the exponent by 1:

$$\frac{dy}{dx} = nk_n x^{(n-1)} + (n-1)k_{(n-1)} x^{(n-2)} + \dots k_1$$

Note that the term k_0 disappears since the corresponding exponent is an implied x^0.

Example: If $y = 3x^2 - 2x + 6$; find $\dfrac{dy}{dx}$

$$\frac{dy}{dx} = 6x - 2$$

Example: If $f(x) = 4x^3 - 5x^2$; find f'(-2).

$$\frac{dy}{dx} = 12x - 10; \text{ substituting } -2 \text{ for x we obtain}$$

$$f'(-2) = 12(-2) - 10 = -34$$

38.4 Integrals of Polynomials

For polynomials, integration is the inverse operation to differentiation. For each term, the exponent is increased by 1 and the term is divided by the new exponent:

$$\int ax^n \, dx = \frac{ax^{(n+1)}}{(n+1)} + C \quad \text{(where C is a constant), with } n+1 \neq 0$$

Example: Evaluate $\int_1^3 x^3 dx = x^4/4 \,]_1^3 = 3^4/4 - 1/4 = 80/4 = 20$

Example: Evaluate $\int_1^3 x^{-2} dx = x^{-1}/-1 \,]_1^3 = 3^{-1}/(-1) - 1/(-1) = \dfrac{-1}{3} + 1 = \dfrac{2}{3}$

Evaluating an integral, as in the two previous examples, gives the area under the curve between the two points. Thus, in the second example, $\dfrac{2}{3}$ is the area under the curve corresponding to the equation x^{-2}, between the points 1 and 3.

So you might be thinking, what about x^{-1}? Does it have an integral? You cannot use the rule above, or you will end up with division by 0. Actually, it does have an integral defined as follows:

$$\int x^{-1} dx = \ln x + C$$

Here, ln x refers to the natural logarithm (or log $_e$) of x.

38.5 Min/Max Values of a Function

I am going to make a general statement, which would require a lot more development to really understand, and a lot more restrictions and qualifications to really be correct. However, it should be good enough to get you through most number sense problems of this type.

Derivative yields slope of tangent line: Let $f(x)$ be a function, and $f'(x)$ represent the derivative of the function, then the slope of a line tangent to $f(x)$ at a point (a,b) is given by $f'(a)$.

Why is the slope of a tangent line important? And what does this have to do with the min/max values of a function?

Consider a simple quadratic equation such as $f(x) = x^2$

The graph of this equation has its vertex at (0,0) and then the left and right sides of the parabola go to infinity. Now picture drawing a tangent (a line that intersects the graph in exactly one point) to this graph where the slope of such a line would be 0 (that is, horizontal). That line would pass through (0,0) – in fact, it would simply be the x-axis.

Now note that if $f(x) = x^2$, then $f'(x) = 2x$. So where is the slope of the tangent line equal to zero? We find this by setting the derivative equal to 0: $2x = 0$, therefore, the slope is equal to 0 at x=0. The value of the function at this point is also 0. It should not take too much to convince you that you have just found the minimum value of this function. We did this by taking the derivative, setting it to 0, solving for x, and substituting this value into the function. Note that this function has no maximum value – it becomes arbitrarily large as x increases.

Note that we would have found the same answer if we had started with $f(x) = -x^2$, except that 0 would be the maximum value, not the minimum. Do you see why?

Example: Find the minimum value of the function $f(x) = 2x^2 - 8x + 21$

Solution: We find the derivative: $f'(x) = 4x - 8$ and set it to 0.

$$4x - 8 = 0$$

As a result, we determine that $x=2$.

Now, we evaluate $f(2) = 8-16+21=13$

Appendix A Conversions (EMH)

A.1 Metric prefixes:

Milli- 1/1000

Centi 1/100

Deci   1/10

Deka-   10

Hector- 100

Kilo-   1000

A.2 common conversions

1 foot	12 inches
1 yard	3 feet
1 mile	5280 feet
1 square mile	640 acres
1 acre	43,560 square feet
1 square foot	144 square inches
1 square yard	9 square feet
1 cubic yard	27 cubic feet
1 pound	16 ounces
1 ton	2000 pounds

1 pint	16 fluid ounces
1 quart	2 pints
1 gallon	4 quarts
1 bushel	4 pecks
1 minute	60 seconds
1 hour	60 minutes
1 square yard	9 square feet
1 mile	1760 yards

x.3 Fahrenheit/Centigrade Conversion

I always had trouble remembering how these conversions go, so I used the two boiling points as the memory key:

$100°\ C = 212°\ F$

So to get from 100 C to 212 F, you multiply by $\dfrac{9}{5}$ and add 32

And from 212 F to 100 C, you subtract 32 and then multiply by $\dfrac{5}{9}$

Appendix B Powers   (EMH)

The more of these powers you have memorized, the better off you are. Study them whenever you get the chance! A typical test will include at least one or two straightforward power problem (e.g., 14^2 or 3^5). However, in several other types of problems it will be helpful to already know these numbers rather than have to compute them on the fly.

x	x^2	x^3	x^4	x^5
Number	Square	Cube	Fourth	Fifth
1	1	1	1	1
2	4	8	16	32
3	9	27	81	243
4	16	64	256	1024
5	25	125	625	3125
6	36	216	1296	7776
7	49	343	2401	16807
8	64	512	4096	32768
9	81	729	6561	
10	100	1000	10000	100000
11	121	1331		
12	144	1728		

13	169	2197		
14	196	2744		
15	225	3375		
16	256	4096		
17	289			
18	324			
19	361			
20	400			
21	441			
22	484			
23	529			
24	576			
25	625			
26	676			
27	729			
28	784			
29	841			
30	900			

MATH IS FUN!

Practice Test 1

Number Sense May 2005

Contestant's Name:_______________________________________

Score: __________

DO NOT UNFOLD THIS SHEET UNTIL TOLD TO BEGIN!

(1) $568 - 677 + 9 =$ _______________

(2) 18% of 220 = _______________

(3) $29^2 =$ _______________

(4) $7 \times 6 + 6 \times 18 =$ _______________

(5) $6\dfrac{2}{3}\% =$ _______________(fraction)

(6) $8 - 11\dfrac{1}{6} =$ _______________

(7) $28 \times 25 =$ _______________

(8) $31 \div 50 =$ _______________(decimal)

(9) $92 =$ _______________(Roman)

(10) $981 + 1620 - 171 =$ _______________

(11) The GCF of 15 and 105 is _______________

(12) $\dfrac{5}{6} - \dfrac{1}{3} + \dfrac{1}{12} =$ _______________

(13) The average of 98, 102, 105, and 95 is _______________

(14) $12.5 \times 120 =$ _______________

(15) 36 is $16\dfrac{2}{3}\%$ of _______________

(16)   37 x 27 = _______________________

(17)   81 –140 –32 + 19 +72 = _____________

(18)   75 x 48 = _________________

(19)   4 square yards = ___________ square ft

(20) $\sqrt{12763}$ = _________________

(21) 9^4 = _________________________

(22)   375 x 168 = _________________

(23)   If 8 dozen eggs cost \$10.56, then 4 eggs will cost \$_______________

(24)   The number of positive integral divisors of 128 is _______________________________

(25)   17 x 19 = _____________________

(26)   The product of the GCF and LCM of x and 36 is 1800; x = _______________

(27)   220 yards = _________ mile (fraction)

(28) $56^2 - 44^2$ = _________________

(29) $7\dfrac{1}{6} x 7\dfrac{5}{6}$ = _________________

(30)   17 x 98 x 6 x 43.1 = _______________

(31) $(12 - 6)(12^2 + (12)(6) + 6^2)$ = _____

(32)   The sum of the roots of $2x^2 - 3x + 7$ = _________________

(33) $41^2 + 15^2 - 26^2$ = _____________

(34) 56^2 = _________________

(35)   26 degrees C = _____________ degrees F

(36)   421 divided by an even prime = _________________

(37) $758 - 857$ = _____________________

(38)   198 x 202 = _____________________

(39)   If x is 11 greater than y, and the product of x and y is 432, then x+y = __________

(40) $\sqrt[3]{3612031}$ = __________________

(41)   143 x 51 x 7 = __________________

(42) 306_8 = _____________________ $_2$

(43)   For a line of equation 17x=323y-15, the slope is ____________________________

(44) $6^{\log_6 5}$ = ___________________

(45) $(6!)^2$ = ___________________

(46)   The radius of circle with area 1444π = _________________________________

(47) i^{122} = ___________________

(48)   87 ½ x 96 = ___________________

(49) 255^2 = ___________________

(50)   1247.32 x 61323 = _________________

(51)   The hypotenuse of an isosceles right triangle is 362. The difference in lengths of the two sides is___________________

(52) $(\sqrt{3} - \sqrt{28})(\sqrt{3} + \sqrt{28})$ = __________

(53) What is the remainder when 392,629 is divided by 9? ___________________

(54) $1+2+3+4+...+99+100 =$ ___________

(55) A triangle has angles with measures 2x, 3x-2, and (x-34); find the square root of x: _______________________________

(56) $2 \sin 22.5° \cos 22.5° =$ _____________

(57) If the area of a triangle is 161.5 sq ft and the base minus the height is 2, what is the height? _____________________

(58) The number of proper subsets of a set with 11 elements is ________________

(59) Find the sum of the coefficients of the binomial expansion of $((x + y)^7 =$ ____________________________

(60) $\pi^7 =$ _________________________

(61) What is the probability of drawing two red cards in a row, without replacement, from a standard 52 card deck? ___________

(62) Find the sum of the 12th and 13th triangular numbers: __________________

(63) If $f(x) = x^2 - 2x + 323$, then $f(17) =$ _______

(64) The product of the roots of $g(x) = ax^3 - bx^2 + 8x - 999$ is 27. Therefore, $a =$ _______________

(65) $\dfrac{8^4}{4^3} = 2^x$, then $x =$ ___________

(66) $\log 40 + \log 25 =$ ___________

(67) $14 \times 69 \times 7 \times 29 =$ _______________

(68) Evaluate: $\lim\limits_{x \to 1} \dfrac{x^2 - 2x - 323}{x - 17} =$ ____

(69) Find the vertical asymptote of $\dfrac{-8x+9}{3x-13}$

(70) $3e^2 + 4\pi =$ _______________

(71) What number most exceeds its square? _________________________________

(72) $\displaystyle\int_1^3 x^{-2}\,dx =$ _____________________

(73) $(17-19i)(17+19i) =$ _________________

(74) Find the 6th pentagonal number: _________

(75) $1 - \tfrac{1}{2} + \tfrac{1}{4} - 1/8 + 1/16 - ... =$ ___________

(76) How many numbers less than or equal to 1000 are relatively prime to 1000? _______

(77) $\dfrac{\tan 29.5° + \tan 15.5°}{1 - \tan 29.5° \tan 15.5°} =$ ___________

(78) $(\sin 60°)^2 =$ ____________

(79) $10! - 9! =$ ____________

(80) $833 \times 16 \times 6 =$ __________

Answer Key to Practice Test 1

Key LISD Invitational

1.   -100
2.   39.6 or 39 3/5
3.   841
4.   150
5.   1/16
6.   -3 1/6 or -19/6
7.   700
8.   .62
9.   XCII
10.   2308-2551
11.   15
12.   7/12
13.   100
14.   1500
15.   216
16.   999
17.   0
18.   3600
19.   36
20.   339-375
21.   6561
22.   63000
23.   .44
24.   8
25.   323
26.   50
27.   1/8
28.   1200
29.   56 5/36
30.   409286-452369
31.   1512
32.   3/2 or 1 ½ or 1.5
33.   1230
34.   3136
35.   78.8 or 78 4/5
36.   210.5 or 210 1/2
37.   -99
38.   39996
39.   22
40.   145-161
41.   51051
42.   11000110
43.   1/19
44.   5
45.   518400
46.   38
47.   -1
48.   8400

49. 65025
50. 72664934 – 80313875
51. 0
52. -25
53. 4
54. 5050
55. 6
56. $\sqrt{2}$ /2
57. 17
58. 2047
59. 128
60. 2869- 3171
61. 25/102
62. 169
63. 0
64. –37
65. 2
66. 3
67. 196098
68. 20
69. 3 ¼ or 13/4
70. 33-36
71. ½
72. 2/3
73. 650
74. 51
75. 2/3
76. 400
77. 1
78. ¼ or .25
79. 51840
80. 75969 - 83966

ABOUT THE AUTHOR

Tom Ferguson holds a Ph.D. in mathematics from the University of Texas-Arlington. He competed at the state level in UIL Number Sense in high school. Tom has over twenty-five years of teaching experience in mathematics at the secondary and college level.